Integrated Circuit Projects

The Maplin series

This book is part of an exciting series developed by Butterworth-Heinemann and Maplin Electronics Plc. Books in the series are practical guides which offer electronic constructors and students clear introductions to key topics. Each book is written and compiled by a leading electronics author.

Other books published in the Maplin series include:

Computer Interfacing	Graham Dixey	0 7506 2123 0
Logic Design	Mike Wharton	0 7506 2122 2
Music Projects	R A Penfold	0 7506 2119 2
Starting Electronics	Keith Brindley	0 7506 2053 6
Audio IC Projects	Maplin	0 7506 2121 4
Auto Electronics Projects	Maplin	0 7506 2296 2
Video and TV Projects	Maplin	0 7506 2297 0
Test Gear & Measurement	Danny Stewart	0 7506 2601 1
Home Security Projects	Maplin	0 7506 2603 8
The Maplin Approach to Professional Audio	T.A. Wilkinson	0 7506 2120 6

Integrated Circuit Projects

Newnes
An imprint of Butterworth-Heinemann Ltd
Linacre House, Jordan Hill, Oxford OX2 8DP

A member of the Reed Elsevier group

OXFORD LONDON BOSTON
MUNICH NEW DELHI SINGAPORE SYDNEY
TOKYO TORONTO WELLINGTON

British Library Cataloguing in Publication Data
A catalogue record for this book is available from the British Library
ISBN 0 7506 2578 3

Library of Congress Cataloguing in Publication Data
A catalogue record for this book is available from the Library of Congress

Edited by Co-publications, Loughborough

Typeset and produced by Sylvester North, Sunderland

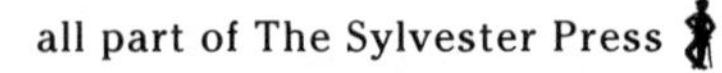
all part of The Sylvester Press

Printed in Great Britain

Integrated Circuit Projects

Newnes
An imprint of Butterworth-Heinemann Ltd
Linacre House, Jordan Hill, Oxford OX2 8DP

A member of the Reed Elsevier group

OXFORD LONDON BOSTON
MUNICH NEW DELHI SINGAPORE SYDNEY
TOKYO TORONTO WELLINGTON

British Library Cataloguing in Publication Data
A catalogue record for this book is available from the British Library
ISBN 0 7506 2578 3

Library of Congress Cataloguing in Publication Data
A catalogue record for this book is available from the Library of Congress

 Edited by Co-publications, Loughborough

 Typeset and produced by Sylvester North, Sunderland

all part of The Sylvester Press 

Printed in Great Britain

Contents

Preface

This book is a collection of projects previously published in *Electronics — The Maplin Magazine*. In their original guise they formed part of the Data File series so popular with regular readers.

Each project is based around an integrated circuit device, selected for publication because of its special features, because it is unusual, because it electronically clever, or simply because we think readers will be interested in it. Some of the devices used are fairly specific in function — in other words, the integrated circuit is designed and built for one purpose alone. Power amplifier integrated circuits are good examples. Others, on the other hand, are not specific at all, and can be used in any number of applications. Bucket brigade delay lines, say, can be used in all sorts of audio projects. Naturally, the circuit or circuits associated with each integrated circuit device reflect this.

While all circuits given with all integrated circuits here are intended for experimental use only — they are not full projects by any means — a printed circuit board track and layout are detailed. To help readers, the printed circuit boards (and kits of parts for some of the projects, too) are available from Maplin, but — as these circuits *are* for experimental use only — all constructional details and any consequent fault finding is left upto readers.

This is just one of the Maplin series of books published by Newnes books covering all aspects of computing and electronics. Others in the series are available from all good bookshops.

Maplin Electronics Plc supplies a wide range of electronics components and other products to private individuals and trade customers. Telephone: (01702) 552911 or write to Maplin Electronics, PO Box 3, Rayleigh, Essex SS6 8LR, for further details of product catalogue and locations of regional stores.

1 Power supply projects

Featuring:

RC4195 ±15 V regulator

The RC4195 is a dual-polarity tracking regulator designed to provide balanced positive and negative 15 V output voltages. This device is designed for local *on-card* regulation, eliminating distribution problems associated with single-point regulation. Only two external components are required for operation (two 10 μF output decoupling capacitors).

Device description

The RC4195 integrated circuit can supply currents of up to 100 mA per supply rail. To keep the device within its maximum power dissipation figure of 600 mW, the maximum input voltage should be 18 V. This is demonstrated by the following equation:

$$\text{Power dissipation per rail} = (V_i - V_o) \times I_L$$

where:

V_i = input voltage
V_o = output voltage
I_L = load current.

This figure is multiplied by 2 (as there are two rails) to give the total integrated circuit power dissipation.

Substituting in the correct figures gives:

$$((18 - 15) \times 0.1) \times 2 = 0.6\,\text{W}$$

A functional block diagram is given in Figure 1.1. Figure 1.2 reveals the internal circuit diagram, while Figure 1.3 shows the integrated circuit pin-out. Tables 1.1 and 1.2

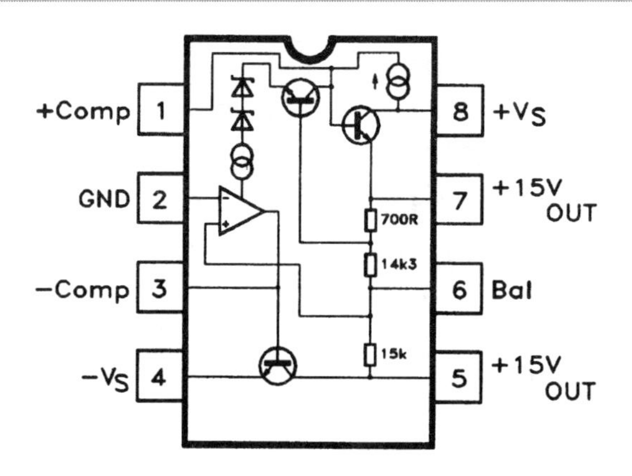

Figure 1.1 Functional block diagram

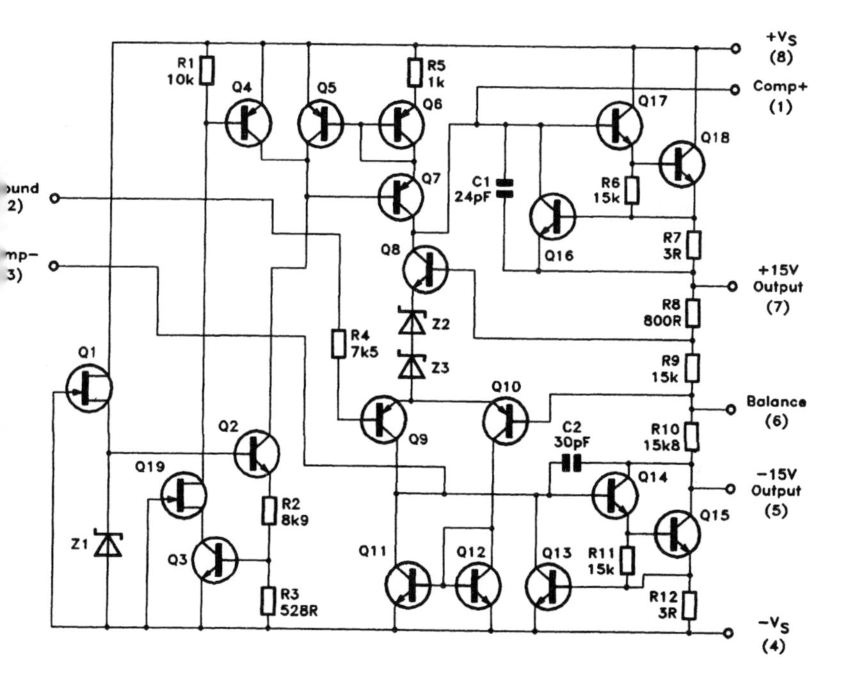

Figure 1.2 IC circuit diagram

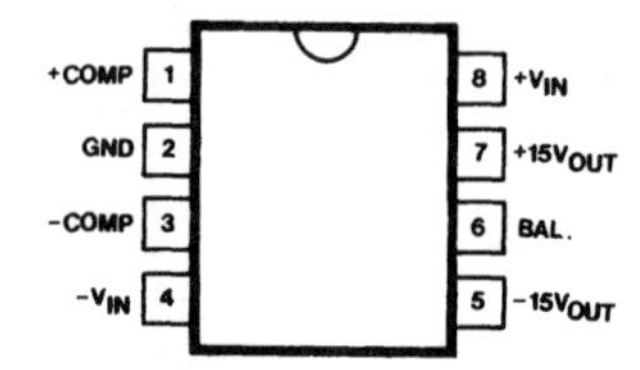

Figure 1.3 4195 IC pin-out

(I_L = ±1m A; V_i = ±20 V; C_L = 10 μF; T_A = 0°C to +70°C)

Parameter	*Conditions*	*Min*	*Typ*	*Max*
Line regulation	V_i = ±18 V to ±30 V		2 mV	20 mV
Load regulation	I_L = 1 mA to 100 mA		5 mV	30 mV
Output voltage drift with temperature			0.005%/°C	0.015%/°C
Supply current	V_i = ±30 V; I_L = 0 mA		±1.5 mA	±4.0 mA
Supply voltage		18 V		30 V
Output voltage	T_j = +25°C	14.5 V	15.0 V	15.5 V
Output voltage tracking			±50 mV	±300 mV
Ripple rejection	f = 120 Hz; T_j = +25°C		75 dB	
Input-output voltage differential	I_L = 50 mA	3 V		
Short circuit current	T_j = +25°C		220 mA	
Output voltage noise	T_j = +25°C; f = 100 Hz to 10 kHz		60 μV rms	
Internal thermal shutdown			175°C	

Note: the specifications given above apply for the given junction temperature, since pulse test conditions are used.

Table 1.1 Electrical characteristics

Maximum junction temperature	125°C
Maximum power dissipation (T_A <50°C)	468 mW
Thermal resistance qJ_A	160°C/W
For T_A >50°C, derate at	6.25 mW per °C

Table 1.2 Thermal characteristics

give the electrical and thermal characteristics respectively, while Table 1.3 gives the absolute maximum ratings. Graphs 1.1 to 1.6 show typical performance characteristics of the device. Figure 1.4 shows a typical application of the regulator, while Figure 1.5 shows the regulator configured to give a single high voltage output. Figure 1.6 shows how to use external pass transistors and current limiting circuitry to increase output current delivery. To balance the output voltages a potentiometer is fitted, with its resistive element connected across the output supply rails and its wiper connected to the *balance* input, as shown in Figure 1.7.

Supply voltage ($\pm V_i$) to ground	±30 V
Load current	100 mA
Operating junction temperature range	0°C to +125°C
Storage temperature range	–65°C to +150°C
Lead soldering temperature (10 sec.)	+300°C

Table 1.3 Absolute maximum ratings

Kit available

A complete kit of parts (including a high quality fibreglass printed circuit board with a component legend to aid component positioning) is available, allowing a basic ±15 V power supply to be constructed using the RC4195 integrated circuit. To aid heat dissipation, it is recommended that the regulator device is soldered directly to the board, so that the copper track acts as a heatsink. The circuit diagram for the kit is given in Figure 1.8, while the printed circuit board legend is shown in Figure 1.9.

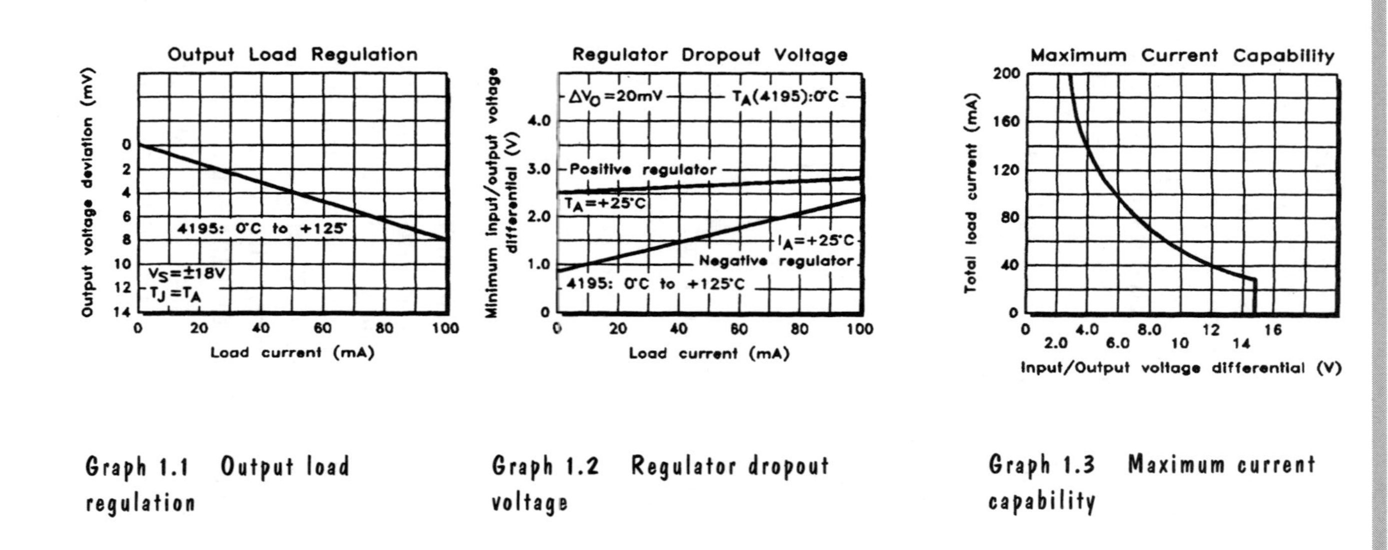

Graph 1.1 Output load regulation

Graph 1.2 Regulator dropout voltage

Graph 1.3 Maximum current capability

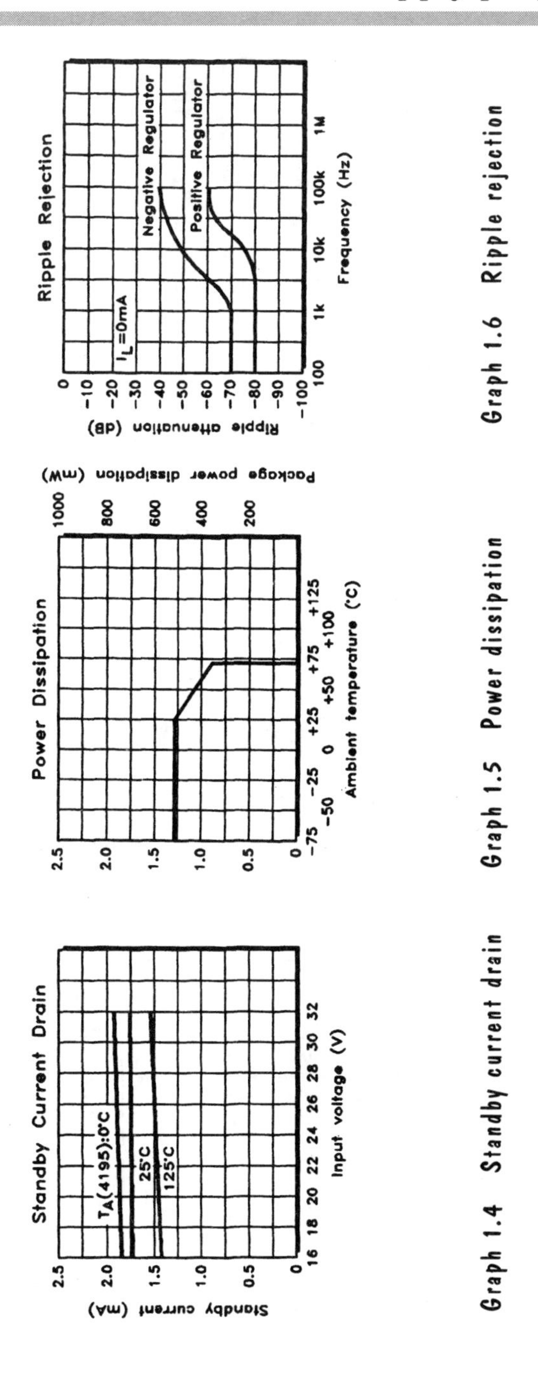

Graph 1.4 Standby current drain

Graph 1.5 Power dissipation

Graph 1.6 Ripple rejection

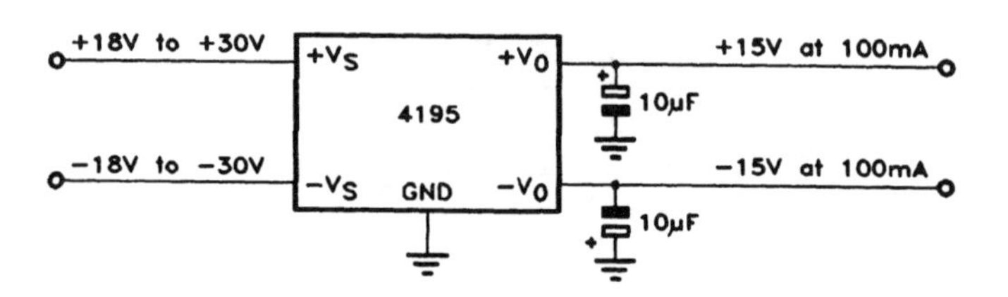

Figure 1.4 4195 set up in balanced output configuration

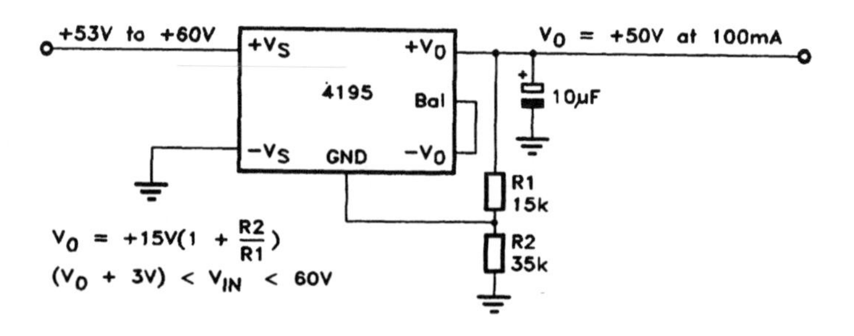

Figure 1.5 Positive single supply (+15 V < V_o < +50 V)

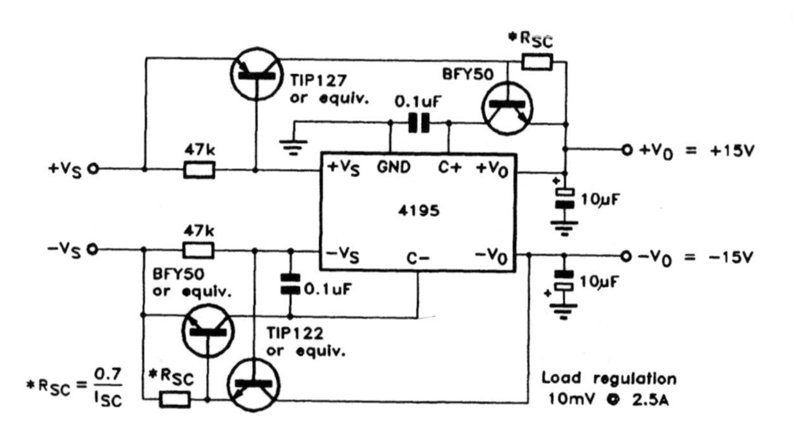

Figure 1.6 4195 set up for high-current output

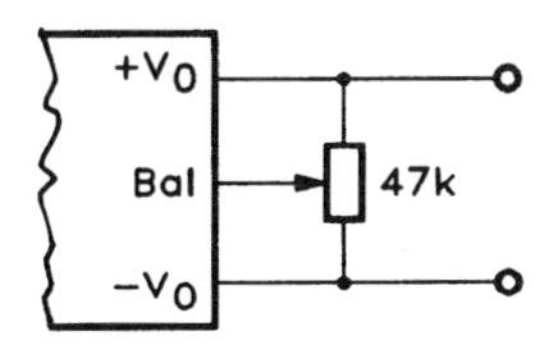

Figure 1.7 Balancing the output rails

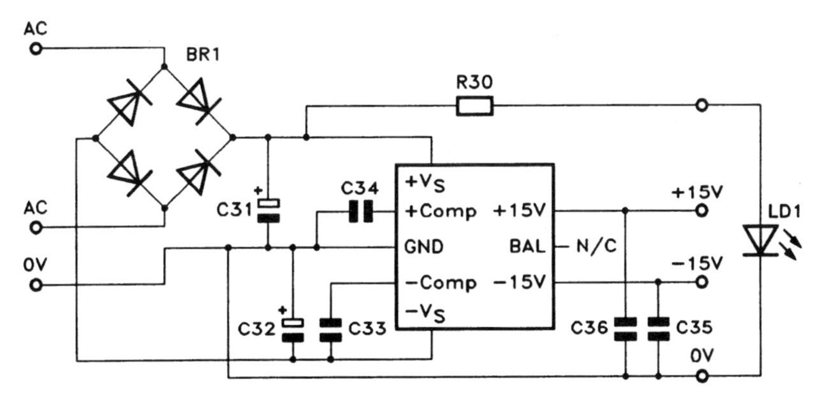

Figure 1.8 Circuit diagram

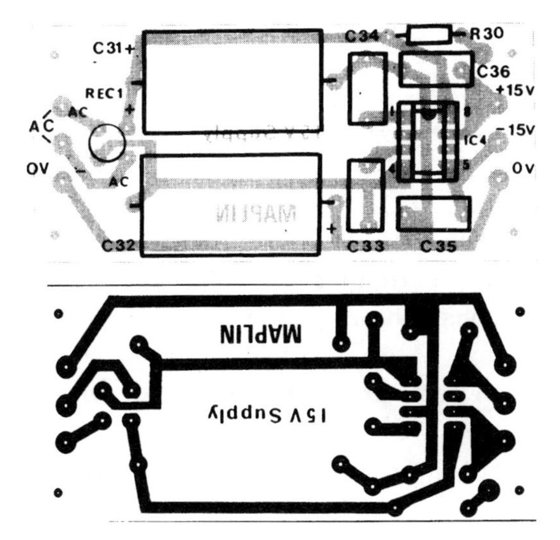

Figure 1.9 PCB legend and track

RC4195 ±15 V regulator parts list

Resistor — 0.6 W 1% metal film

R30	1k8	1	(MIK8)

Capacitors

C31,32	1000 µF 35 V PC elect	2	(FFI8U)
C33,34, 35,36	100 nF polyester	4	(BX76H)

Semiconductors

IC4	4195	1	(XX02C)
BR1	W01	1	(QL38R)
LD1	LED red	1	(WL27E)

Miscellaneous

	pins 2141	1	(FL21X)
	PCB	1	(XX04E)
	instruction leaflet	1	(XT33L)
	constructors' guide	1	(XH79L)

The above items are available as a kit, order as LP88V

L200 adjustable voltage current regulator

The L200 is a monolithic IC designed for programmable voltage and current regulation. Voltage outputs between 2.85 V and 36 V may be accommodated, at currents of up to 2 A. The device is supplied in a 5-pin package; the IC pin-out is shown in Figure 1.10. The L200 has internal protection to minimise the possibility of damage to the device; this comprises current limiting, power limiting, thermal shutdown and input over-voltage protection (up to 60 V for 10 ms). Table 1.4 shows typical electrical characteristics for the device. In addition some typical performance figures are shown in Figure 1.11.

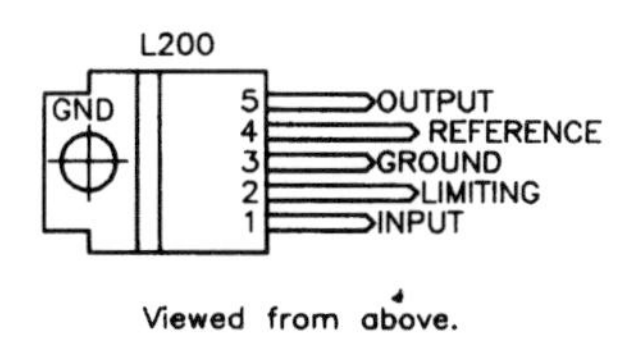

Figure 1.10 L200 IC pin-out

Parameter	*Conditions*	*Min*	*Typ*	*Max*
D.C. input voltage	Absolute maximum			40 V
Quiescent current drain (pin 3)	Input voltage (V_i) = 20 V		4.2 mA	9.2 mA
Output voltage range	Output current = 10 mA	2.85 V		36 V
Operating junction temperature range (L200C)	Absolute maximum	–25°C		+150°C
Line regulation	V_i = 8 V to 18 V, V_o = 5 V	48 dB	60 dB	
Dropout voltage between pins 1 and 5	Output current = 1.5 A $\Delta V_o \Delta$ 2%		2 V	2.5 V
Reference voltage (pin 4)	Input voltage (V_i) = 20 V Output current (I_o) = 10 mA	2.64 V	2.77 V	2.86 V

Table 1.4 L200 typical electrical characteristics

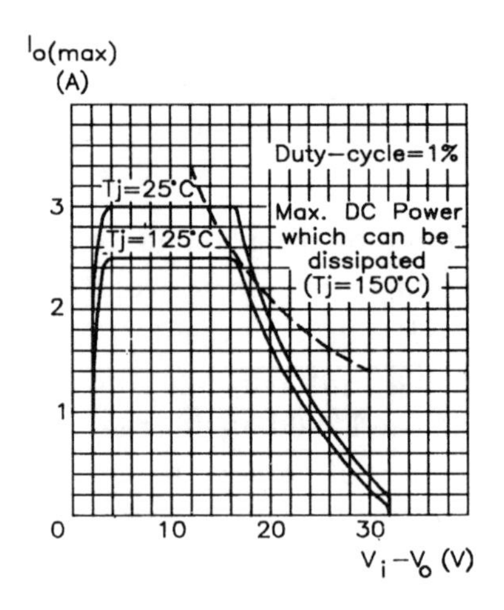

Figure 1.11(a) Typical safe operating area protection

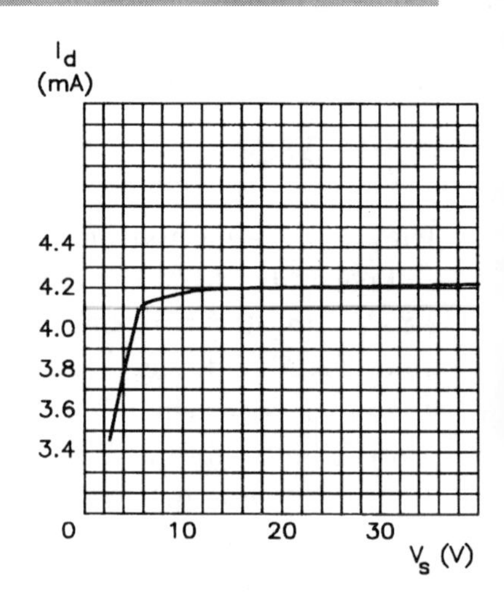

Figure 1.11(b) Quiscent current (I_d) vs, voltage (V_s)

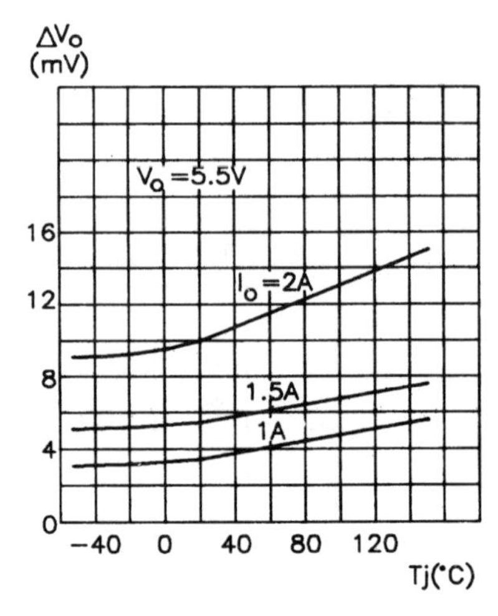

Figure 1.11(e) Voltage load regulation (ΔV_o) vs junction temperature (T_j)

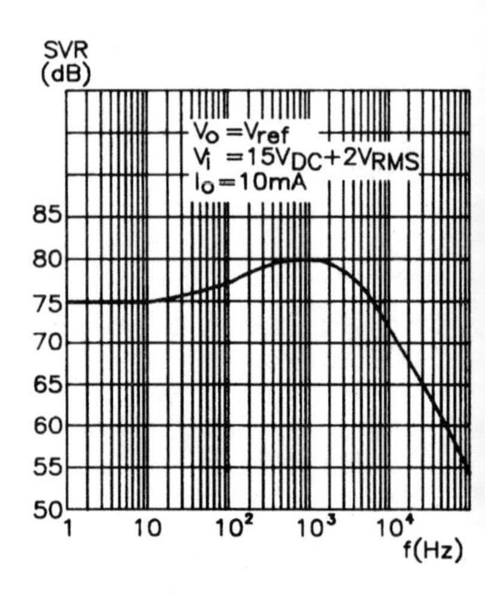

Figure 1.11(f) Supply voltage rejection (SVR) vs frequency (f)

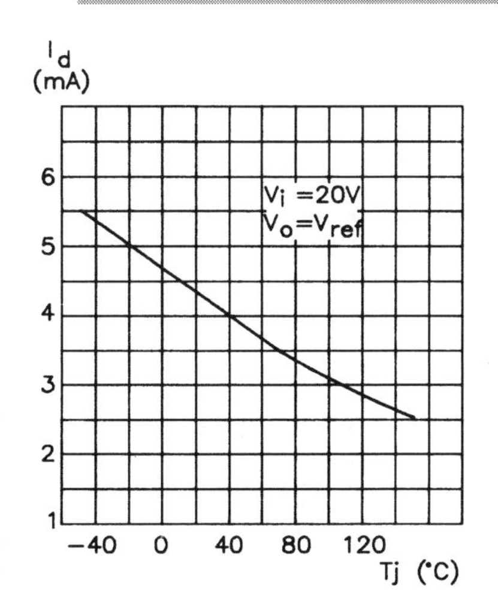

Figure 1.11(c) Quiescent current (I_d) vs junction temperature (T_j)

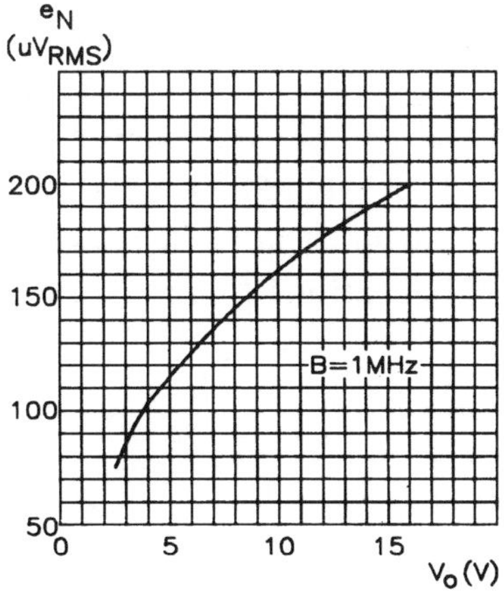

Figure 1.11(d) Output noise voltage (°N) vs output voltage (V_o) for 1 MHz bandwidth

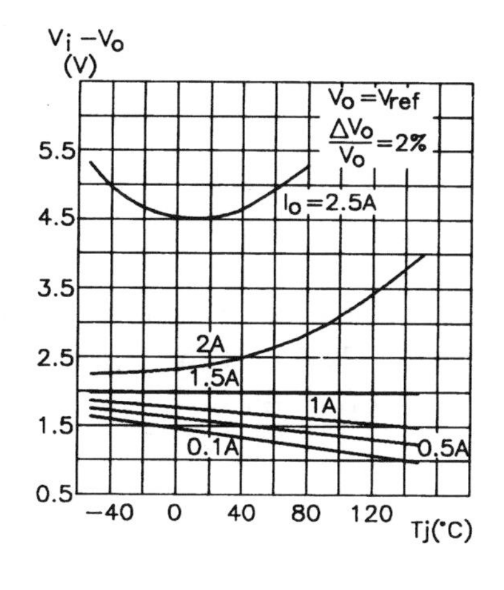

Figure 1.11(g) Dropout voltage ($V_i - V_o$) vs junction temperature (T_j)

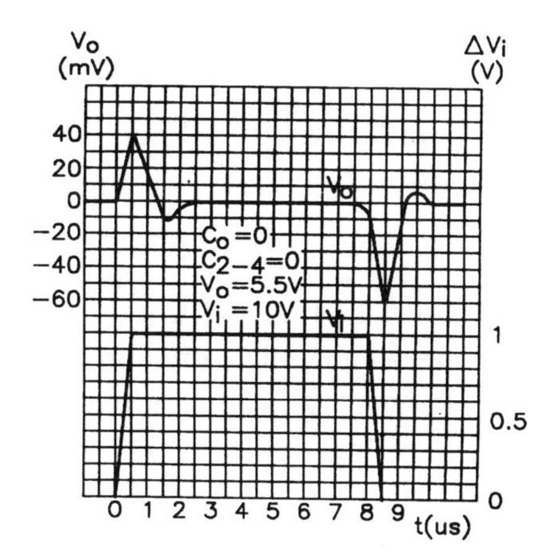

Figure 1.11(h) Voltage transient response

General description

As can be seen from the block diagram shown in Figure 1.12, the L200 regulator uses a relatively sophisticated design. The device may be used in several different configurations to provide voltage or current regulation.

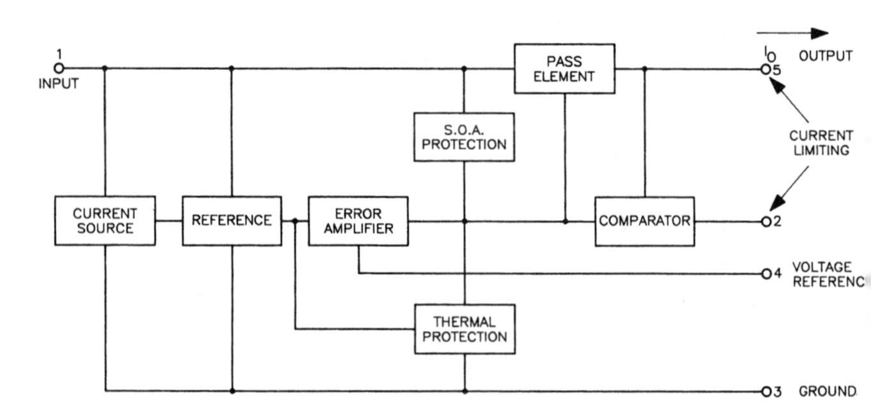

Figure 1.12 L200 block diagram

Current limiting is controlled by connecting a resistor between pin 2 and pin 5 of the L200. The current limit threshold is approximated by the expression:

$$I_O = V_{LC} \div R_{SC}$$

where:

I_O = output current (A)

V_{SC} = current limit sense voltage (V)

R_{SC} = resistance between pin 5 and pin 2 (Ω).

The current limit sense voltage is variable depending on several factors including load and temperature but is typically 0.45 V.

Power dissipation is controlled by the internal Safe Operating Area (SOA) protection circuitry of the L200. The device can supply a current of up to 2 A as long as the input/output differential voltage is less than 20 V. With differential voltages above 20 V the maximum current output drops considerably; if this value is exceeded, then the SOA protection limits the output current so as to reduce power dissipation and prevent damage to the device.

Output voltage is determined by the values of the resistors connected between pin 3 and pin 4, and pin 4 and pin 2 of the device. The final output voltage may be approximated by the expression:

$$V_O = V_{ref} (1 + (R2 \div R1))$$

where:

V_O = output voltage (V)

V_{ref} = reference voltage on pin 4 (V)

R1 = resistance between pin 4 and pin 3 (Ω)

R2 = resistance between pin 2 and pin 4 (Ω).

For the purpose of approximate calculation, V_{ref} may be taken at a typical value of 2.77 V although in practice this figure may vary very slightly.

Kit available

A kit of parts is available for a basic application circuit using the L200 regulator IC. The kit includes a high quality fibreglass PCB with a printed legend to aid component positioning. Figure 1.13 shows the circuit diagram of the module and Figure 1.14 shows the legend. To allow the module to be as versatile as possible, some of the component positions on the PCB are left open, so that the parameters of the module may be determined by the user. In particular, the values of resistors R2 and R3 (which determine the current limiting threshold of the regulator) are subject to selection, depending on the individual application.

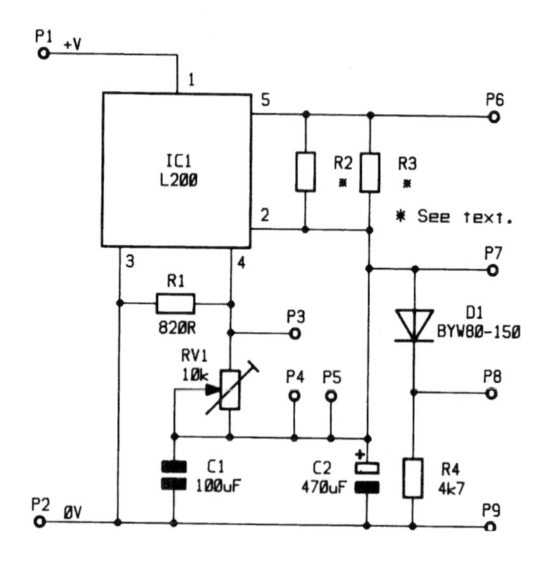

Figure 1.13 Module circuit diagram

For connection information reference should be made to the wiring diagram shown in Figure 1.15. Input connections to the module are made to P1(Input +V) and P2(0 V). Output connections are made to P8(Output +V) and P9(0 V).

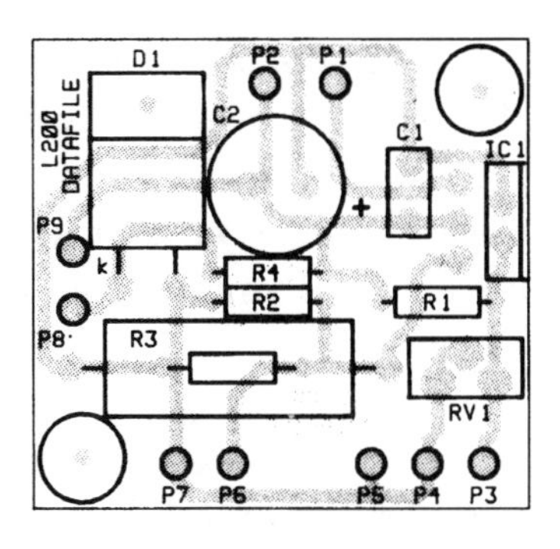

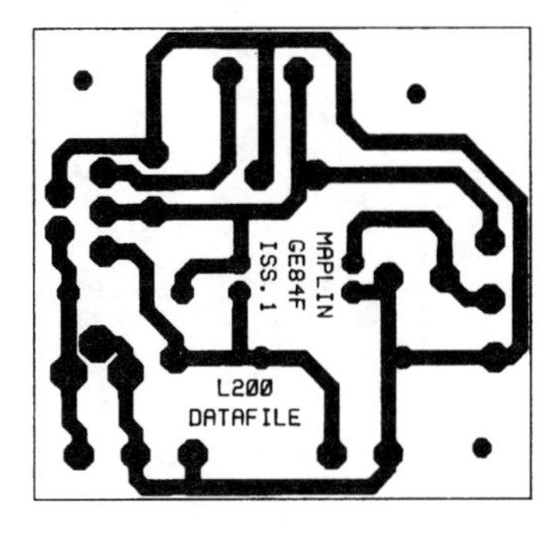

Figure 1.14 PCB legend and track

To maintain correct regulation, it is important that the input/output differential voltage is never allowed to fall below the regulator dropout voltage. The dropout voltage may vary but as a general rule, it is recommended that the input voltage is always at least 4 V above the maximum required output voltage.

Output voltage control is via preset resistor RV1. Provision is also made for an external voltage control potentiometer and this may be connected to P3, P4 and

P5. If an external voltage control is used, then RV1 should *not* be fitted. If a fixed output is required, then a fixed resistor may be connected between P3 and P4; once again, RV1 should *not* be fitted as this is effectively in parallel with any external voltage control resistors.

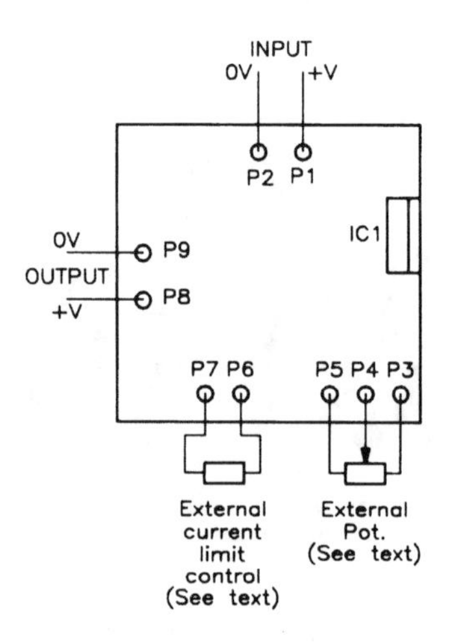

Figure 1.15 Wiring diagram

The current limit threshold of the module is set by resistors, R2 and R3. Two parallel resistors are used to enable the very low values of resistance required for higher current limit thresholds to be achieved. It should be noted that R3 may either be a 0.6 W or a 3 W type and a separate set of holes is provided for both types. For some applications it may of course be possible to achieve the correct value using one resistor only. Provision is also included for an external current limiting resistor, which may be connected between P6 and P7. A low value variable resistor may be used for variable current limit control but at higher current levels the resolution will become increasingly poor. The approximate current limit threshold may be calculated using the following method, assuming a typical voltage of 0.45 V between P6 and P7 (pin 5 and pin 2 of the IC):

$$I_O = 0.45 \div R_{LC}$$

where I_0 is the output current and R_{SC} is the total parallel resistance between P6 and P7 (R2, R3 and any external current limit resistor in parallel) in ohms.

Heatsink

At higher power levels, it is necessary to use a suitable heatsink to prevent IC1 from reaching excessive temperatures. The type of heatsink used is dependent on the individual application. In some cases, a large area of metal such as the side of an enclosure may already be available. The tab of the L200 is at 0 V potential and will bolt directly to a heatsink if this is also at 0 V potential; however, in some cases it may be necessary to isolate the tab of the L200 (if the heatsink is not at 0 V potential). An insulating bush and a greaseless or mica washer should be used for this purpose, as illustrated in Figure 1.16.

Typical heatsinks for use with the L200 up to 20 W are shown in Table 1.5. The parameters shown are intended to provide general guidelines and the power ratings may be found to vary slightly in different applications.

Please note: any of the higher power heatsinks are also suitable for lower power applications and where the power dissipation is variable, the maximum power dissipation under worst case conditions should be used for the purposes of selecting a heatsink.

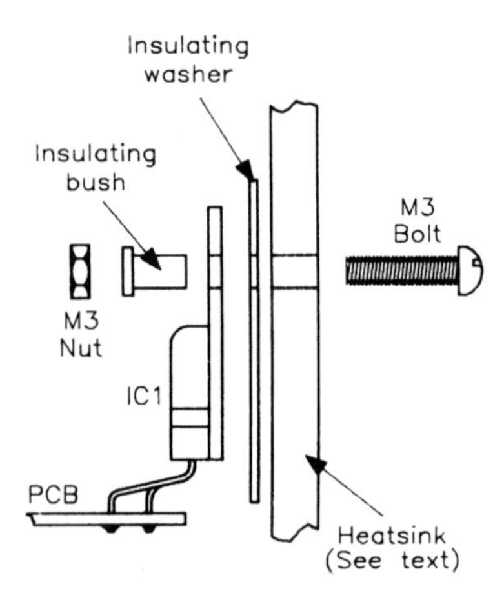

Figure 1.16 Heatsinking

Regulator power dissipation	*Heatsink (stock code)*
Up to 500 mW	No heatsink required
500 mW–1.5 W	Vaned heatsink TO126 (JX21X)
1.5 W–3.5 W	High power twisted vane (FG55K)
3.5 W–10 W	Heatsink 4Y (FL41U)
10 W–20 W	Flat heatsink (FL42V)

Table 1.5 Typical heatsinks for the L200

In addition to heatsinking for the L200, it is also recommended that a small heatsink (such as JW21X) is used for D1 when the module is used at current levels in excess of 750 mA; in this case D1 should be mounted vertically and the heatsink bolted to the tab.

Table 1.6 shows the specification of the prototype module.

Parameter	*Conditions*	
Input voltage		5 V–35 V
Output voltage		2.8 V–32.8 V
Output current (max)	For input/output differential voltage less than 20 V	2 A
Quiescent current (max)		11 mA

Table 1.6 Specification of prototype module

L200 voltage/current regulator parts list

Resistors — All 0.6 W 1% metal film (unless specified)

R1	820 Ω	1 (M820R)
R2	see text	1
R3	see text	1
R4	4k7	1 (M4K7)
RV1	10 k vert encl preset	1 (UH16S)

Capacitors

C1	100 nF monores cap	1 (RA49D)
C2	470 μF 35 V PC elect	1 (FF16S)

Semiconductors

IC1	L200	1 (YY74R)
D1	BYW80–150	1 (UK63T)

Miscellaneous

P1–9	pins 2145	1 (FL24B)
	L200 PCB	1 (GE84F)
	instruction leaflet	1 (XT00A)
	constructors' guide	1 (XH79L)

The above items are available as a kit, order code LP69A

78xxx/79xxx series voltage regulators

Around 20 years ago, the first three-terminal integrated circuit voltage regulators were introduced — and they revolutionised the design of low-to-mid current fixed-voltage d.c. power supplies. Circuits which would have previously used a large handful of components in their design could then be made up with relatively fewer (the mains transformer, bridge rectifier, smoothing and decoupling capacitors are all that are essentially required in addition to the regulator integrated circuit itself). As a result, the use of one of these devices could save considerable space and money — and so, spurred on by demand from industry, the range of regulator types flourished (to include both positive [78xxx] and negative [79xxx] types) while costs were reduced as the result of huge production volumes.

Of this huge number of different regulators, most component catalogues — the Maplin catalogue being no exeption — often list over 20 or so. Typical voltage regulators are featured in Table 1.7, along with the other components that you will need to build a complete power supply unit, based around a particular regulator. These components include Maplin-designed printed circuit boards which greatly simplify the construction of power supplies based around these devices. 5 V, 12 V or 15 V power supplies can be designed with positive or negative outputs. Group 1 of the table lists 100 mA regulators, Group 2 lists regulators rated between 500 mA and 1 A, and Group 3 features devices capable of supplying currents of up to 2 A.

Fixed voltage regulators: ±5 V to ±15 V, 100 mAto 2 A

Group 1

Regulator/ Stock Code	*Voltage*	*Current*	*Transformer/ Stock Code*	*PCB/ Stock Code*	*Rectifier/ Stock Code*
LM78LO5ACZ	+5V	100mA	6–0–6 VAC	Yes	WO-005
QL26D			WBOOA	YQ39N	QL37S
LM79L05ACZ	–5V	100mA	6–0–6 VAC	Yes	WO-005
WQ85G			WBOOA	YQ39N	QL37S
LM78L12ACZ	+12V	100mA	9–0–9 VAC	Yes	WO-005
WQ77J			WB01B	YQ39N	QL375
LM79L12ACZ	–12V	100mA	9–0–9 VAC	Yes	WO-005
WQ86T			WB01B	YQ39N	QL37S
LM78L15ACZ	+15V	100mA	12–0–12 VAC	Yes	WO-005
QL27E			WB02C	YQ39N	QL37S
LM79L15ACZ	–15V	100mA	12–0–12 VAC	Yes	WO-005
WQ87U			WB02C	YQ39N	QL37S

All transformer secondaries connected in series, see Figure 1.17a

Additional Parts List for Group 1

			Nut M2.5
Pins 2141	1 Pkt	FL21X	Shake Washer M2.5
Heatsink 92F	1	HQ79L	Solder Tag M2.5
Screw M2-5 x 10mm	1 Pkt	JY30H	Red LED Low Current

The above is in addition to the selected regulator, transformer, PCB, dropper resistor.

Group 2

Regulator/ Stock Code	*Voltage*	*Current*	*Transformer/ Stock Code*	*PCB/ Stock Code*	*Rectifier/ Stock Code*
L78M05CV	+5V	500mA	0–6, 0–6 VAC	Yes	WO-005
QL28F			WBO6G	YQ40T	QL37S
LM79M05CT	–5V	500mA	0–6, 0–6 VAC	Yes	WO-005
WQ88V			WBO6G	YQ41 U	QL37S
L78M12CV	+12V	500mA	0–9, 0–9 VAC	Yes	WO-005
QL29G			WB11M	YQ40T	QL37S
LM79M12CT	–12V	500mA	0–9, 0–9 VAC	Yes	WO-005
WQ89W			WB11M	YQ41U	QL37S

Table 1.7 Regulated d.c. power supply options, based around

R1 or Link/ Stock Code	C1/ Stock Code	C2/ Stock Code	C3/ Stock Code	LED Resistor/ Stock Code
Link	220μF 35V	100nF Cer	-	620Ω
–	JL22Y	BX03D	-	M620R
Link	220μF 35V	100nF Cer	-	620Ω
–	JL22Y	BX03D	-	M620R
Link	220μF 35V	100nF Cer	-	2k
–	JL22Y	BX03D	-	M2K
Link	220μF 35V	100nF Cer	-	2k
–	JL22Y	BX03D	-	M2K
47Ω	220μF 35V	100nF Cer	-	2k7
W47R	JL22Y	BX03D	-	M2K7
47Ω	220μF 35V	100nF Cer	-	2k7
W47R	JL22Y	BXO3D	-	M2K7

Inclusion of an LED is necessary for minimum load.

1 Pkt	JD62S
1 Pkt	BF45Y
1 Pkt	LR65V
1	UK48C

bridge rectifier, R1 where applicable, capacitors C1–2 and LED

R1 or Link/ Stock Code	C1/ Stock Code	C2/ Stock Code	C3/ Stock Code	LED Resistor/ Stock Code
-	1000μF 35V	10μF 50V	100nF Cer	620Ω
-	FF18U	FFO4E	BXO3D	M620R
-	1000μF 35V	10μF 50V	100nF Cer	620Ω
-	FF18U	FFO4E	BXO3D	M620R
-	1000,μF 35V	10μF 50V	100nF Cer	2k
-	FF18U	FFO4E	BXO3D	M2K
-	1000μF 35V	10μF 50V	100nF Cer	2k
-	FF18U	FFO4E	BXO3D	M2K

ready-made Maplin PCBs **(Continued pages 26, 27 and 28)**

L78M15CV	+15V	500mA	0–9, 0–9 VAC	Yes	WO-005
QL30H			WB11M	YQ40T	QL37S
LM79M15CT	–15V	500mA	0–9, 0–9 VAC	Yes	WO-005
WQ90X			WB11M	YQ41U	QL37S
L7805CV	+5V	1A	0–6, 0–6 VAC	Yes	WO-005
QL31J			YJ50E	YQ40T	QL37S
L7905CV	–5V	1A	0–6, 0–6 VAC	Yes	WO-005
WQ92A			YJ50E	YQ41U	QL37S
L7812CV	+12V	1A	0–12, 0–12VAC	Yes	WO-005
QL32K			WB25C	YQ40T	QL37S
L7912CV	–12V	1A	0–12, 0–12VAC	Yes	WO-005
WQ93B			WB25C	YQ41U	QL37S
L7815CV	+15V	1A	0–12, 0–12VAC	Yes	WO-005
QL33L			WB25C	YQ40T	QL37S
L7915CV	–15V	1A	0–12, 0–12VAC	Yes	WO-005
QL36P			WB25C	YQ41U	QL37S

All transformer secondaries connected in series, see Figure 1.17a

Group 3

Regulator/ Stock Code	*Voltage*	*Current*	*Transformer/ Stock Code*	*PCB/ Stock Code*	*Rectifier/ Stock Code*
L78S05CV	+5V	2A	0–6, 0–6 VAC	Yes	S005
UJ54J			YJ51F	YQ40T	QLO9K
L78S09CV	+ 9V	2A	0–6, 0–6 VAC	Yes	S005
UJ55K			YJ51F	YQ40T	QLO9K
L78S12CV	+12V	2A	0–15, 0–15 VAC	Yes	S005
UJ56L			WB12N	YQ40T	QLO9K
L78S15CV	+15V	2A	0–20, 0–20 VAC	Yes	S005
UJ57M			WB12N	YQ40T	QLO9K

For transformer YJ51F, connect secondaries in series, see Figure
for +12V or 1.20b for +15V. Inclusion of an LED is necessary for

Additional Parts List for Groups 2 and 3

Item	Quantity	Stock Code
8W Hi-Fi Heatsink	1	HQ81C
Pin2141	1 Pkt	FL21X
Screw M2.5 x 10mm	1 Pkt	JY30H
Nut M2-5	1 Pkt	JD62S
Shake Washer M2.5	1 Pkt	BF45Y
Red LED Low Current	1	UK48C

The above is in addition to the selected regulator, transformer, PCB,

Table 1.7 (Continued)

-	1000μF 35V	10~F 50V	100n F Cer	2k7
-	FF18U	FFO4E	BXO3D	M2K7
-	1000μF 35V	10μF 50V	100nF Cer	2k7
-	FF18U	FFO4E	BXO3D	M2K7
-	2200μF 35V	10μF 50V	100nF Cer	620Ω
-	JL28F	FFO4E	BXO3D	M620R
-	2200,μF 35V	10μF 50V	100nF Cer	620Ω
-	JL28F	FFO4E	BXO3D	M620R
-	2200μF35V	10μF50V	100nFCer	2k
-	JL28F	FFO4E	BXO3D	M2K
-	2200μF35V	10μF50V	100nFCer	2k
-	JL28F	FFO4E	BXO3D	M2K
-	2200μF35V	10μF50V	100nFCer	2k7
-	JL28F	FFO4E	BXO3D	M2K7
-	2200μF35V	10μF50V	100nFCer	2k7
	JL28F	FFO4E	BXO3D	M2K7

Inclusion of an LED is necessary for minimum load.

R1 or Link/ Stock Code	*C1! Stock Code*	*C2/ Stock Code*	*C3! Stock Code*	*LED Resistor/ Stock Code*
-	4700μF 35V	10μF 50V	100nF Cer	620
-	JL30H	FFO4E	BXO3D	M620R
-	4700μF 35V	10μF 50V	100nF Cer	1k5
-	JL30H	FFO4E	BXO3D	M1 K5
-	4700~F35V	10μF50V	100nFCer	2k
-	JL30H	FFO4E	BXO3D	M2K
-	4700μF 35V	10μF 50V	100nF Cer	2k7
-	JL30H	FFO4E	BXO3D	M2K7

1.19a; for WB12N, connect in parallel, as shown in Figure 1.20a minimum load.

bridge rectifier, capacitors C1-3 and LED dropper resistor.

Further Notes

Many of the regulators require additional heatsinking. The following shows how the size of the required heatsink can be found mathematically; however the calculations do not take into account the thermal resistance between the package of the device and the heatsink surface, and the package dissipation. Hence the use of heatsink compound at the jointing faces will ensure maximum thermal conduction and is highly recommended (Maplin code HQOOA for a small syringe).

The power dissipation formula is: power dissipation = d.c. input voltage – d.c. output voltage × maximum output current for example: 30 V – 5 V x 1 A = 25 W and to find a suitable size of heatsink, the maximum safe power dissipation level of the device can be subtracted, leaving the remainder which must be taken care of by the heatsink, as follows: max. device dissipation = 8 W (for example) 25 – 8 = 17 W

Then the ambient environment temperature must be taken into consideration: max. safe package temperature = 80°C – room temperature 25°C = 55°C, 55 / 17 = 3.235°C per watt

Thus the heatsink chosen must be one with a maximum temperature rise in the centre of 3°C per watt.

Table 1.7 (Continued)

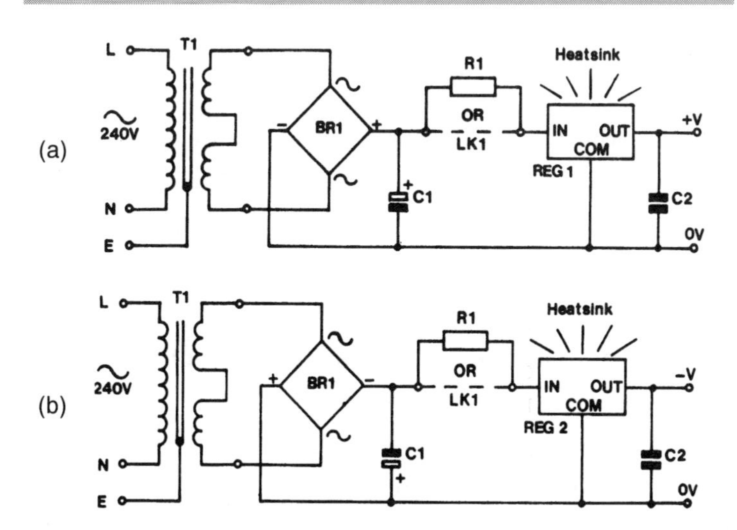

Figure 1.17 (a) Circuit diagram for 100 mA positive regulators, (b) circuit diagram for 100 mA negative regulators (see Group 1 of Table 1.7)

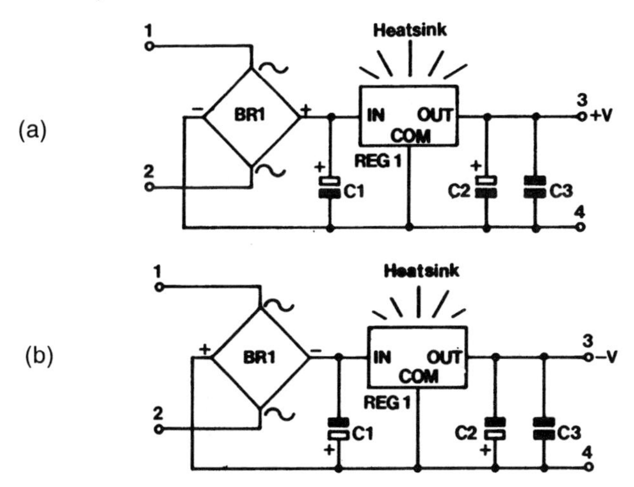

Figure 1.18 Circuit diagram for positive regulators rated between 500 mA and 2 A, (b) circuit diagram for negative regulators rated between 500 mA and 2 A (see Groups 2 and 3 of Table 1.7)

Explanation of specifications

Some of the specifications given in Table 1.8 are self-explanatory, while some need clarifying:

- output current is the absolute maximum peak value permissible through the regulator before the current-limiting protection circuit operates,
- output voltage is the regulated output voltage for which the stated specifications apply,
- line regulation is the ratio of change in output voltage to the change in input voltage under constant load conditions. This measurement was made with a small load (low power dissipation), or by using pulse techniques, so that the temperature of the chip die is not affected. This parameter is expressed as a percentage of the output voltage,

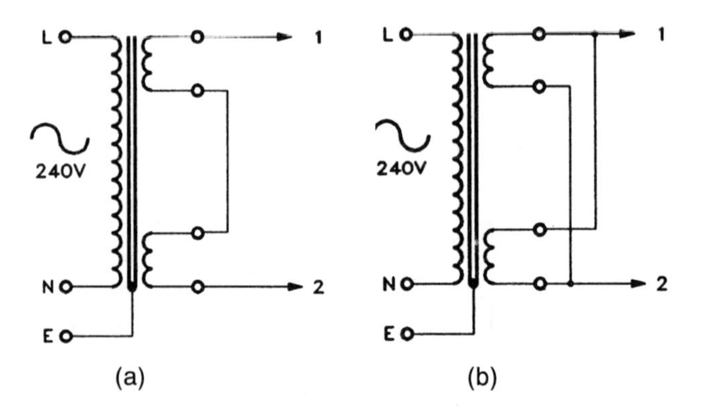

Figure 1.19 (a) Transformer connections for series windings, (b) transformer connections for parallel windings

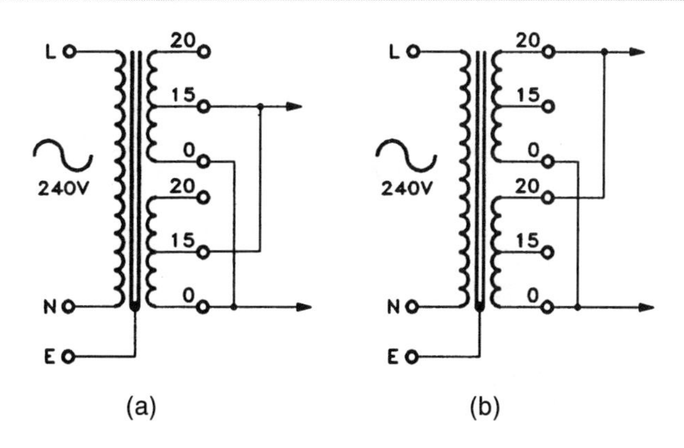

Figure 1.20 (a) Parallel connections for multi-tap transformer (for use with 12 V regulator of Groups 3 in Table 1.7), (b) parallel connections for multi-tap transformer (for use with 15 V, 2 A regulator of Group 3 in Table 1.7)

- load regulation is the ratio of change in output voltage to the change in load current, again measured at a constant die temperature. This factor is also expressed as a percentage of the output voltage,
- ripple rejection is similar to line regulation, but is specified at the rectifier frequency (full wave), and is expressed in dB,
- quiescent current is the current drawn by the regulator with no load attached, expressed in mA,
- input voltage range is the minimum-to-maximum working input voltage to the regulator,

78/79 voltage regulators

Order	Type No.	Output Current (max)	Output Voltage (typ)	Line Regulation (typ)	Load Regulation (typ)
QL26D	LM78LO5ACZ	100mA	+5V±4%	0.36%	0.4%
WQ77J	LM78L12ACZ	100mA	+12V±4%	0.25%	0.25%
QL27E	LM78L15ACZ	100mA	+15V±4%	0.25%	0.25%
QL28F	L78MO5CV	500mA	+5V±4%	0.06%	0.4%
QL29G	L78M12CV	500mA	+ 2V ±4%	0.07%	0.2%
QL30H	L78M15CV	500mA	+15V±4%	0.07%	0.17%
QL31J	L7805CV	1A	+5V±4%	0.06%	0.2%
QL32K	L7812CV	1A	+12V±4%	0.085%	0.07%
QL33L	L7815CV	1A	+15V±4%	0.075%	0.055%
UJ54J	L78SO5CV	2A	+5V±4%	100mVmax	80mVmax
UJ55K	L78SO9CV	2A	+9V±4%	130mVmax	100mVmax
UJ56L	L78S12CV	2A	+12V±4%	240mVmax	150mVmax
UJ57M	L78S15CV	2A	+15V±4%	300mVmax	150mVmax
WQ85G	LM79LO5ACZ	100rnA	5V±5%	1%	0,2%
WQ86T	LM79L12ACZ	100mA	12V±5%	1%	0.2%
WQ87U	LM79L15ACZ	100mA	–15V±5%	1.5%	0.3%
WQ88V	LM9MO5CT	500mA	–5V±4%	0.14%	1.5%
WQ89W	LM79M12CT	500mA	–12V±4%	0.075%	0.55%
WQ90X	LM79M15CT	500mA	–15V±4%	0.06%	0.45%
WQ92A	L7905CV	1A	–5V±4%	0.06%	0.2%
WQ93B	L7912CV	1A	–12V±4%	0.085%	0.07%
QL36P	L7915CV	1A	–15V ±4%	0.075%	0.055%

Table 1.8 Electrical and physical characteristics of 78xxx/79xxx

℩ipple ℩ejection dB)(typ)	Quiescent Current (typ)	Input Voltage Range	Output Resistance	Output Noise Voltage	Short Circuit Current	Case Style
62dB	3mA	7V to 30V	0.2Ω	40μV	-	TO92r
54dB	3mA	14,5V to 35V	0.2Ω	80μV	-	TO92r
51dB	3.lmA	17.5V to 35V	0.2Ω	90μLV	—	TO92r
80dB	4.5mA	7V to 25V	0.05Ω	40μV	300mA	P1d
80dB	4.8mA	14.5V to 30V	0.05Ω	75μV	240mA	P1d
70dB	4.8mA	17.5V to 30V	0.05Ω	90μV	240mA	P1d
78dB	42mA	7V to 25V	0.017Ω	40μV	750mA	P1d
71dB	4.3mA	14.5V to 30V	0.0181Ω	75μV	350MA	P1d
70dB	4.4mA	17.5V to 30V	0.019Ω	90μV	230mA	P1d
54dB	8mA	8V to 35V	0.017Ω	40μV	500mA	P1d
47dB	8mA	12V to 35V	0.017Ω	60μV	500mA	P1d
47dB	8mA	15V to 35V	0.018mΩ	75μV	500mA	P1d
46dB	8mA	18V to 35V	0.019mΩ	90μV	500mA	P1d
60dB	3mA	7V to –25V		40μV	-	TO92n
55dB	3mA	14.5V to –35V		80μV	-	TO92n
52dB	3mA	–17.5V to –35V		90μV	-	TO92n
60dB	1mA	–7V to –25V		125μV	140mA	P1n
60dB	1.5mA	–14.5V to –30V		300μV	140mA	P1n
59dB	1.5ma	–17.5V to –30V		375μV	140mA	P1n
60dB	1 mA	–7V to –25V	-	125μV	750mA	P1n
60dB	1.5mA	–14.5V to –30V	-	300μV	350mA	P1n
60dB	1.5mA	–17.5V to –30V	-	375μV	230mA	P1n

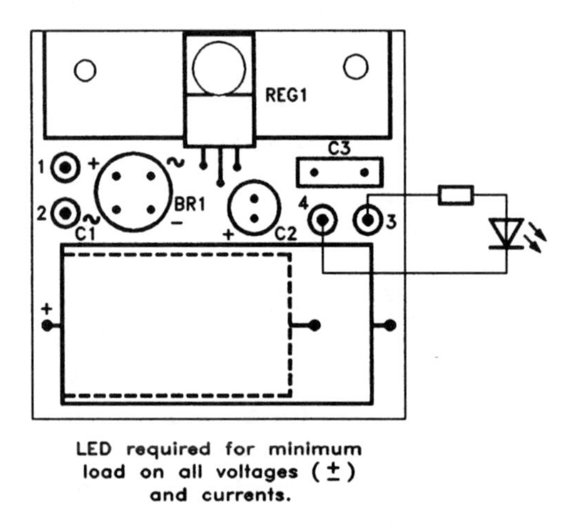

Figure 1.21 YQ40T PCB, intended for use with fixed-voltage positive regulators rated at between 500 mA and 2 A (see Groups 2 and 3 of Table 1.7

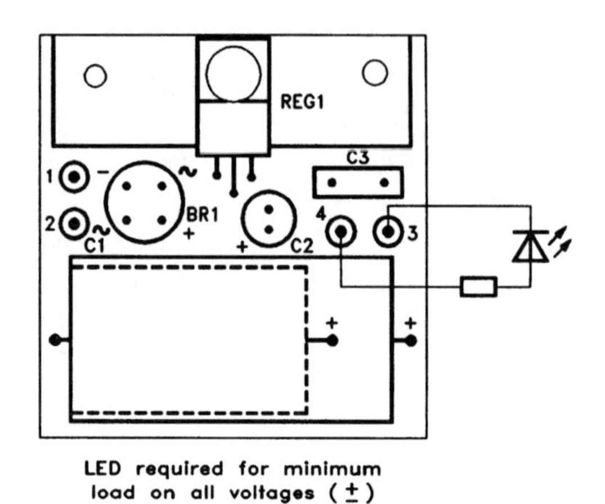

Figure 1.22 YQ41U PCB, intended for use with fixed-voltage negative regulators rated at between 500 mA and 2 A (see Groups 2 and 3 of Table 1.7)

- output impedance expresses the phase relationship (reactance) between voltage and current at the output of the device. The major contributory factor to a regulator's output impedance is the collector-emitter resistance of the IC's internal pass transistor. It is measured in Ω or mΩ,

- output noise voltage is the r.m.s. a.c. voltage at the output, with a constant load and no input ripple, measured over a specified frequency range,

- short-circuit current is the maximum current available from the regulator with the output shorted to ground. This value is lower than the maximum output current, due to the current fold-back circuitry.

Function of the capacitors

The function of capacitor C1 (all regulator printed circuit boards) is to store enough energy between the rectifier pulses to provide a smooth supply rail with minimal ripple (under full load conditions), thus preventing regulator *drop out*. The latter condition occurs when the supply rail falls below the regulator's minimum input voltage.

The purpose of capacitor C2 (500 mA to 2 A printed circuit boards only) is to provide decoupling, and prevent instability in the regulator. A value of 10 μF is sufficient for most applications; however, the value can be increased in circumstances where the peak current of the load circuit exceeds the maximum current of the regulator.

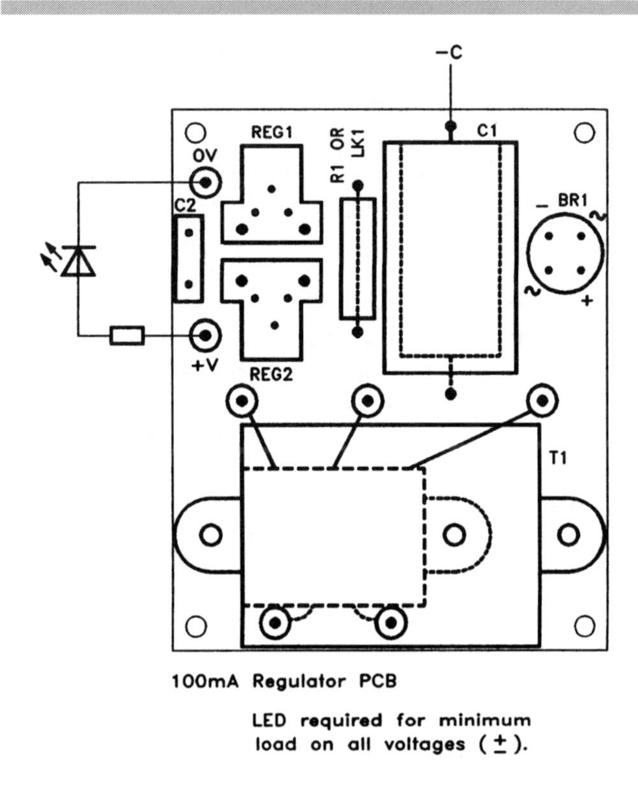

Figure 1.23 YQ39N PCB shown populated for use with 100 mA fixed-voltage positive regulators (see Group 1 of Table 1.7)

The ceramic capacitor C3 (C2 on 100 mA printed circuit board) provides high-frequency decoupling, which aids stability.

General comments

- the average output ripple on 2 A regulators is 10 mV peak-to-peak; for all others it is 5 mV peak-to-peak,

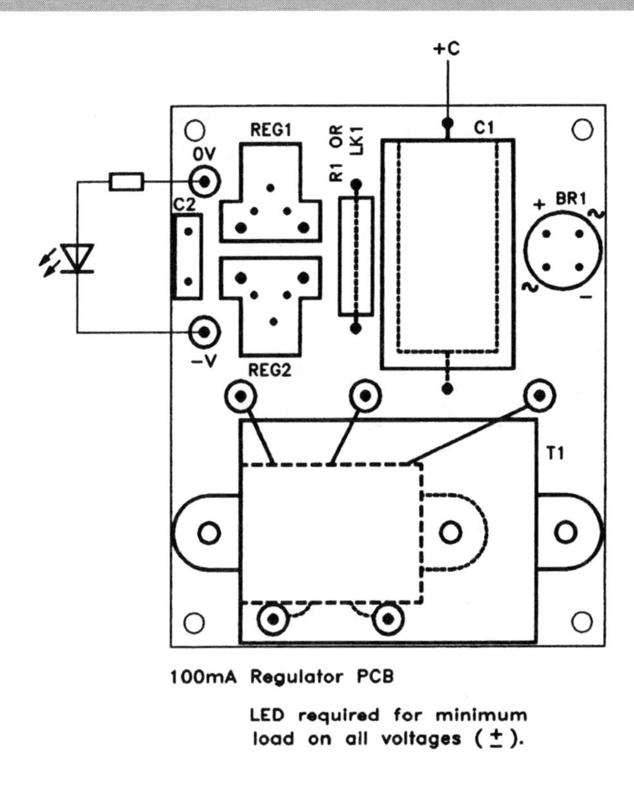

Figure 1.24 YQ39N PCB, shown populated for use with 100 mA fixed-voltage negative regulators (see Group 1 of Table 1.7)

- use M2.5 hardware for attaching the regulator to the heatsink,
- please note that the mounting tab of all regulators is an electrical connection, and this fact should be borne in mind when mounting the device on a heatsink.

A mounting kit (e.g. WR23A or QY45Y/JR78K) should therefore always be used, unless:

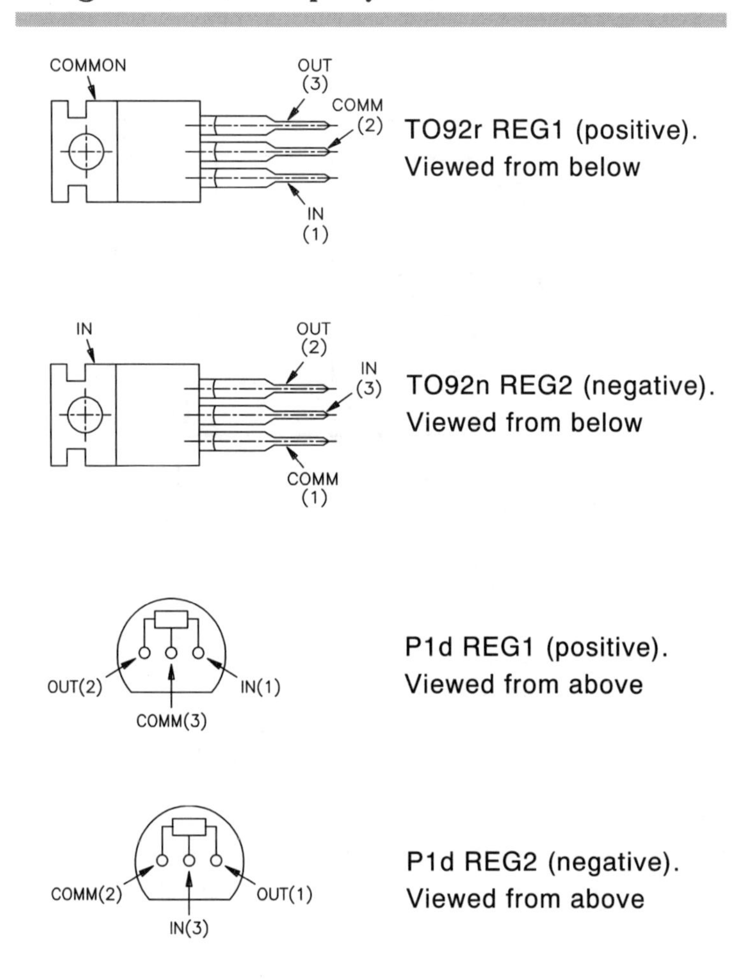

Figure 1.25 Package styles and pin-outs for the regulators. The TO92 r/n case styles are used for the 100 mA LM7xLxxACZ devices, listed in Group 1 of Table 1.7. The P1d/n case styles are used for the others, which are featured in Groups 2 and 3 of Table 1.7. Note that the pin-outs of positive and negative devices are different — be sure to place the device in the correct position on the PCB. REG1 denotes a positive device, while REG2 denotes a negative version — as shown in Figures 1.17 and 1.18

(a) in the case of negative voltage regulators, the heatsink is at the same potential as the input voltage,

(b) in the case of positive voltage regulators, the heatsink is at 0 V potential,

(c) the heatsink is completely electrically isolated.

Unless greaseless insulating washers are used, remember to use heat transfer compound (Maplin stock numbers: HQ00A or FL79L), to aid thermal conductivity between the integrated circuit and the heatsink.

- to provide power-on indication, a standard 5 mm LED should be connected between the output of the regulator and ground, via a series resistor of suitable value (see Table 1.7 for values). Such an arrangement is compulsory for negative regulators — this is to prevent the off-load output voltage decreasing excessively (i.e. going *more* negative).

Load regulation

To get the best performance from a regulator (assuming that adequate supply decoupling is used) a couple of important factors must be borne in mind, namely power dissipation and differential voltage.

The differential voltage can be found by subtracting the specified minimum output voltage from the minimum input voltage. This differential voltage must be maintained at all times — otherwise regulator *drop- out* (loss of regulation) may occur. Worst-case values must always be assumed for a variable power supply.

Power dissipation of the regulator can be calculated by multiplying the differential voltage by the current. The maximum power dissipation of a TO220 package is approximately 2 watts without heatsink, and around 18 watts with an infinite heatsink. In comparison, a TO92 package can only dissipate around half a watt without a heatsink.

To keep the heatsink to a minimum size, and obtain maximum current from the regulator, the input-to-output voltage must be kept to a minimum (without the regulator *dropping out* during the power supply's ripple troughs). This will in turn reduce power dissipation, and therefore the heat output. It may therefore be necessary to use a transformer with a higher VA rating than the minimum requirement to improve the load regulation.

Some general notes on heatsinks

Basic heatsink calculation

Heatsinks, which can be found in the *Semiconductors* section in the Maplin Catalogue, are rated universally in terms of temperature rise per watt of power dissipated (°C/W). The lower the figure, the smaller the temperature rise will be from a given heat energy source (e.g. transistor or regulator IC), and the more reliable the electronic system will be. However, the trade-off for a heatsink with a lower thermal rating is increased bulkiness and cost, and so it is good practice to determine the optimum heatsink for a particular application, rather than just to choose one which could well be over-specified.

A very quick *rule of thumb* calculation can be made by dividing the maximum allowable temperature rise by the power dissipation of the device and adding 10% (i.e. multiply by 1.1) to the total (for safety, as it is a simple calculation):

$$\frac{T_{MAX} - T_A}{V_{DIFF} \times I} \times 1.1 = TR$$

Where:

T_{MAX} = maximum allowable temperature (°C)

T_A = ambient temperature (°C)

V_D = differential voltage

T_R = thermal rating of heatsink, expressed in °C/W

Example:

$$\frac{60°C - 25°C}{8V \times 1A} \times 1.1$$

$= 3.977°C/W$ (physically larger, or smaller °C/W number)

In most applications this type of calculation is all that is necessary, but in some circumstances a more accurate calculation is required to include the thermal resistance, Δ (theta), of the device and mounting arrangement.

Thermal resistance is a useful concept as it has broadly similar properties to electrical resistance and can be modelled on Ohm's law, where heat flow is analogous to current, and temperature is the equivalent of voltage. This can be seen from the *formula triangle* shown below.

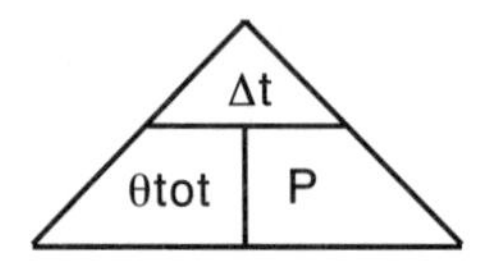

where:

Δt = change in temperature,

t = (max t – ambient t)

P = heat dissipation (watts)

θtot = °C/W = t/P

Series thermal resistances are added together in the same way as conventional electrical resistors:

$$\theta tot = \theta 1 + \theta 2 + \theta 3 + \ldots + \theta n$$

Thermal resistance in parallel is calculated in the usual manner of *sum over product*, or by using the reciprocal method:

$$\frac{\theta 1 + \theta 2}{\theta 1 \times \theta 2} = \theta tot \text{ or } \frac{1}{\theta 1} + \frac{1}{\theta 2} + \frac{1}{\theta 3} = \frac{1}{\theta tot}$$

Typical thermal resistances that will be encountered:

θjc = junction to case

θcs = case to (heat)sink

θsa = (heat)sink to air

θtot = total thermal resistance

These thermal ratings can generally be found in manufacturers' data.

Other more complex thermal calculations (e.g. for air flow/forced cooling) are really beyond the scope of this book.

2 Remote control projects

Featuring:

UM3750 encoder/decoder

The UM3750 encoder/decoder is a digital code transmitter-receiver system. Table 2.1 gives the electrical characteristics for this device and Figures 2.1 and Table 2.2 show the pin-out and pin designation respectively. Working in the transmission mode, the UM3750 will sequentially encode and transmit the twelve bits of data presented to it in parallel on pins one to twelve. Each of the twelve inputs may be a logic 0 or 1 allowing a total of 4096 unique codes. On-chip pull-up resistors are provided so that simple, single pole switches may be used to set the transmitted code. The rate of transmission is set by an external RC network running at a frequency of approximately $^{2}/_{RC}$ The actual format of the output code is shown in Figure 2.2. It can be seen from this, that with a 100 kHz clock (R = 100 kΩ, C = 180 pF), it takes 11.52 ms to transmit one word. Each word is separated by a 11.52 ms space giving a total transmission time of 23.04 ms (43.4 Hz). In the receive mode, the incoming

Parameter	*Conditions*	*Minimum*	*Maximum*
Operating voltage		3.0 V	11 V
Input voltage *low*		V_{ss}	V_{ss} + 0.5 V
Input voltage *high*		V_{DD} $^{-0.5\,V}$	V_{DD}
Output logic level *low*	I_{sink} = 2 mA	Vs_s	V_{ss} + 1 V
Output logic level *high*	I_{source} = 5 μA	V_{DD} – 0.5 V	V_{DD}

Note: above specifications assume a 9 V power supply and an ambient temperature of 25°C.

Table 2.1 Electrical characteristics of the UM3750

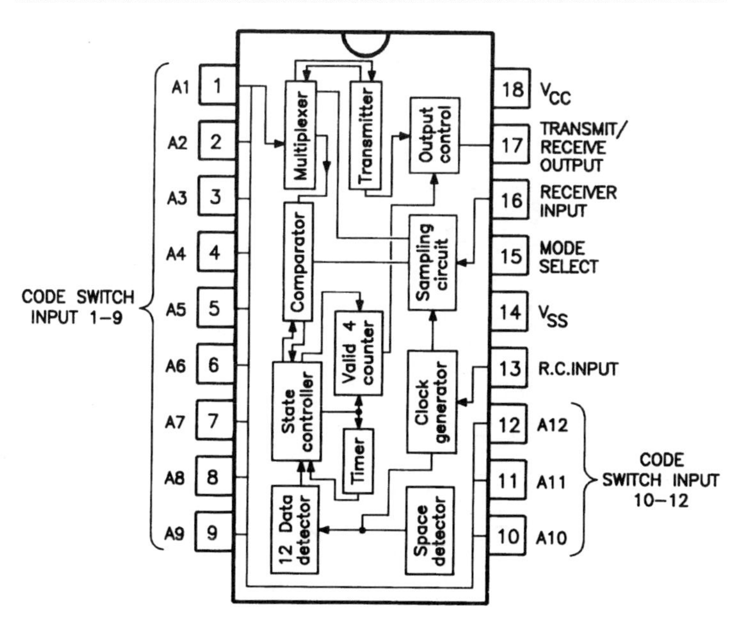

Figure 2.1 Pin-out of the UM3750

Pin	*Designation*	*Description*
1–12	A1–A12	These data select lines are used to set the address of the encoder/decoder pair. They have on-chip pull-up resistors.
13	RC input	RC input pin for single pin oscillator. A resistor is hooked from this pin to (Vs_s). Frequency = $^2/_{RC}$.
14	V_{ss}	The ground pin of the UM3750.
15	Mode select	This pin changes the IC from receiver mode (grounding this pin) to transmitter mode (taking this pin to V_{cc}).
16	Receiver input	The receiver's input for the digital PCM waveform.
17	Transmit/receive output	In the transmitter mode this pin is the PCM output. In the receiver mode this pin is active low to signal a valid received code.
18	V_{cc}	The positive supply pin.

Table 2.2 Pin designation of the UM3750

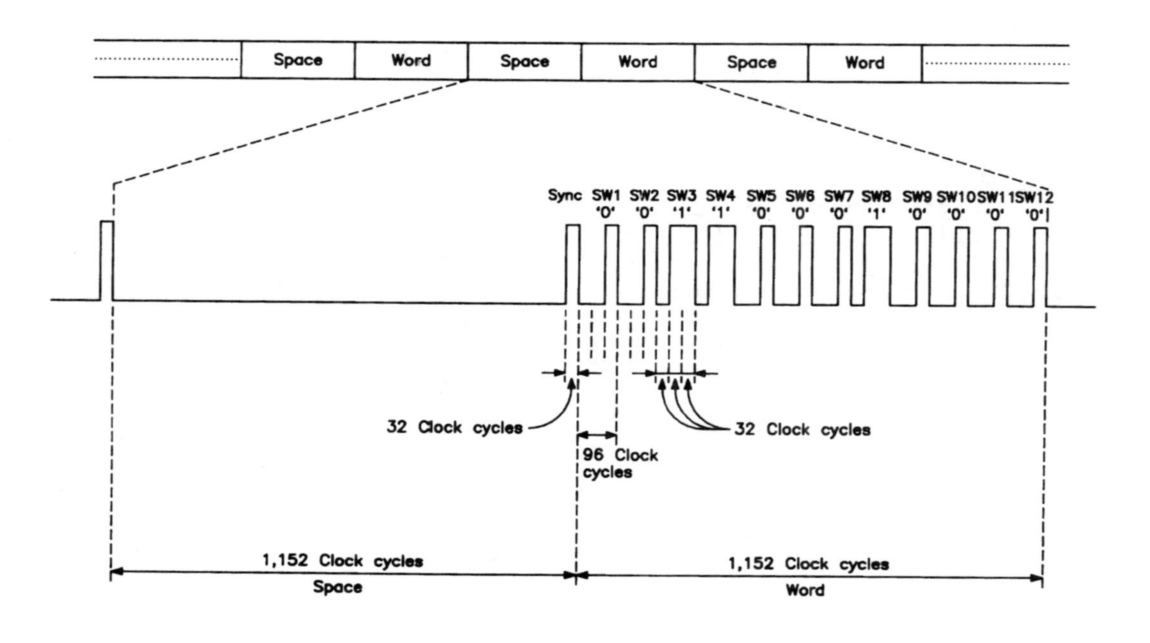

Figure 2.2 Output waveform of the UM3750

signal is compared with the local code in a sequential manner. As soon as an error is detected the system resets and restarts comparison on the next word.

If all 12 bits are received correctly a three stage counter is triggered and after receiving a further 3 consecutive valid codes, the receive output pin goes low. After this pin goes low, one in five codes are required to be valid to maintain this condition (one valid code has to be received in 11520 clock pulses or the receive output changes back to its high state).

Using the UM3750

The basic circuit configuration for the device is shown in Figure 2.3. Up to 4096 receivers can be used at one time by setting a unique code at each unit. If a large number of receivers are driven from the same source it is important to ensure that the transmission line provides sufficient drive capability. The frequency of the transmit and receive clocks need to be matched to within ±50% to allow for the use of multiple T_x/R_x links without any fear of crosstalk.

A high quality fibreglass PCB is available allowing both an infra-red transmitter and receiver to be constructed on the same board, but not at the same time! Figure 2.4 shows the combined circuit diagram that was used to produce the PCB, the track layout of which is shown in Figure 2.5. Both the receive and transmit circuits have provision for a voltage regulator, zener diode, and resistor (RG1, ZD1 and R5) to allow operation at supply voltages up to 35 volts.

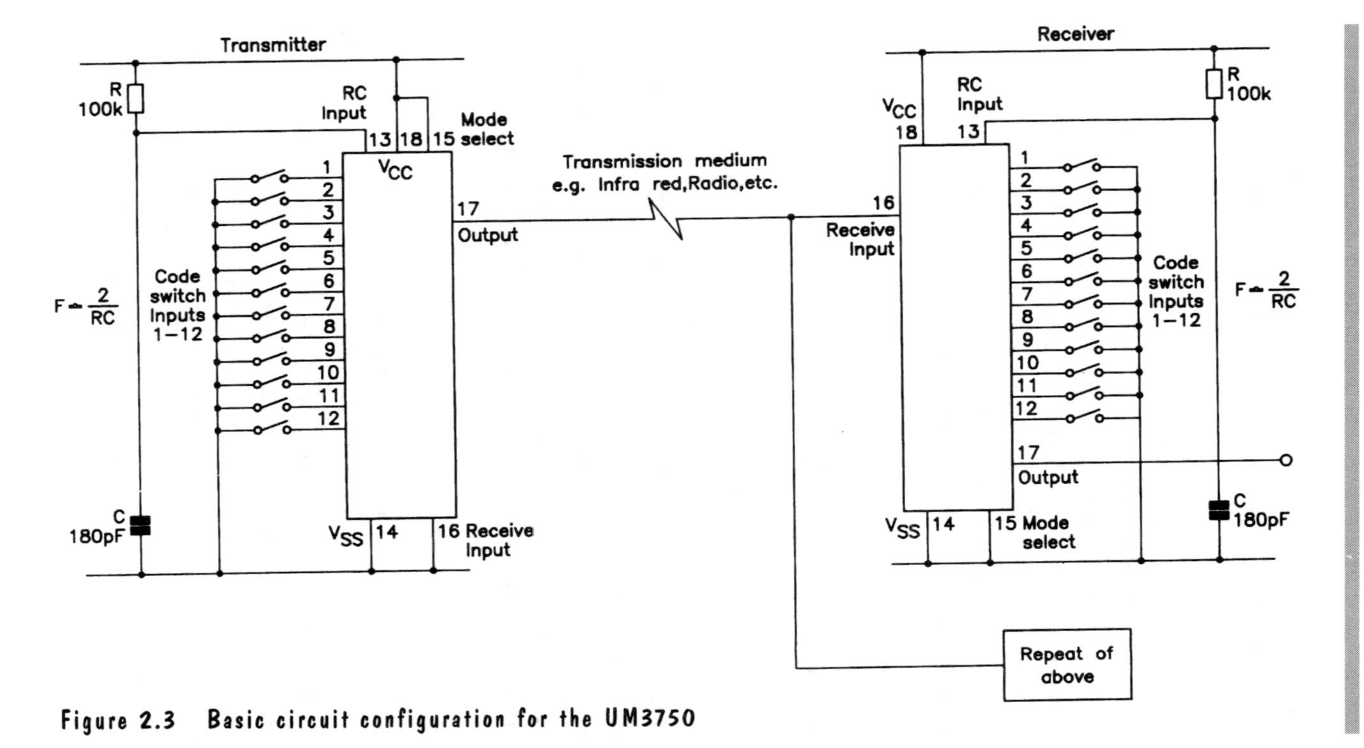

Figure 2.3 Basic circuit configuration for the UM3750

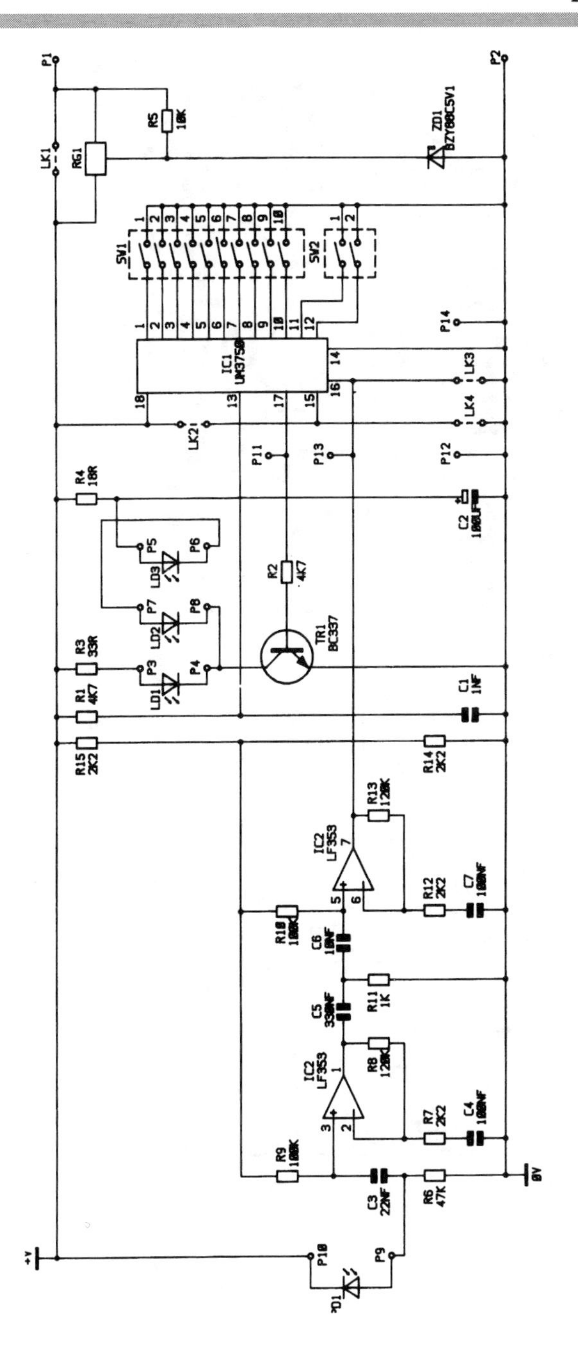

Figure 2.4 Combined circuit to which the PCB is designed

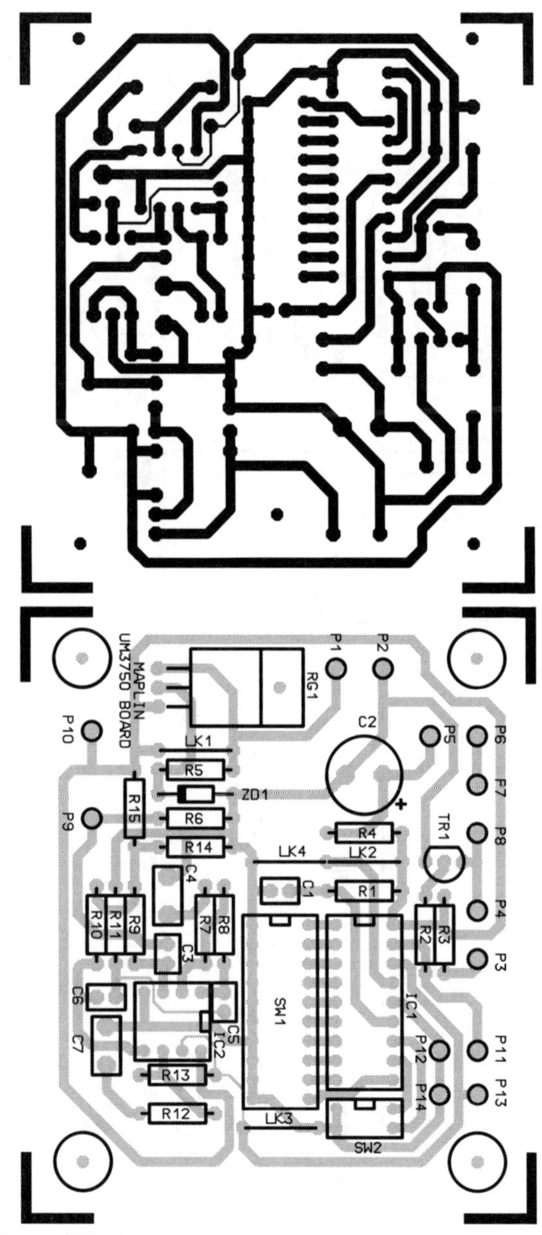

Figure 2.5 PCB layout

Figure 2.6 shows the circuit diagram for the transmitter and Figure 2.7 shows the PCB wiring. The circuit shown is designed to operate at a supply voltage of 9 volts. For supply voltages below 9 volts, R3 and R4 need to be changed to the values shown in Table 2.3.

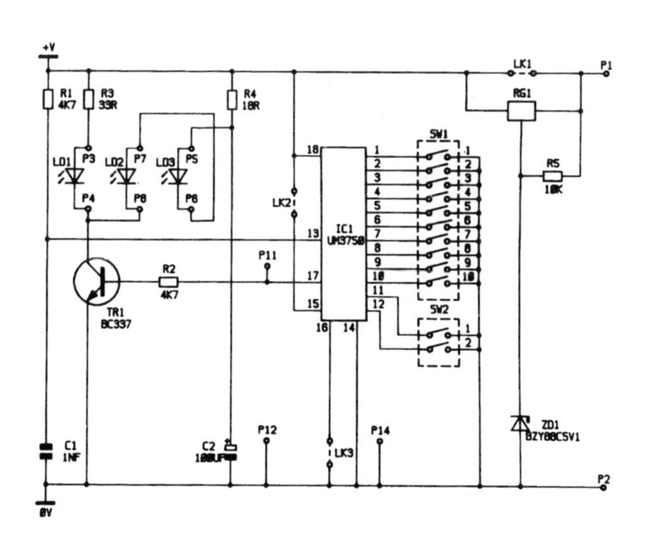

Figure 2.6 Transmitter circuit diagram

Power supply	*Comments*	*R3*	*R4*
6 V	Fit LK1	18 Ω	8.2 Ω
9 V	Fit LK1	33 Ω	18 Ω
12–35 V	Fit RG1, ZDI, R5	39 Ω	22 Ω

Max Frequency ($F = {}^{2}/_{R1C1}$)

Table 2.3 Selecting R3 and R4

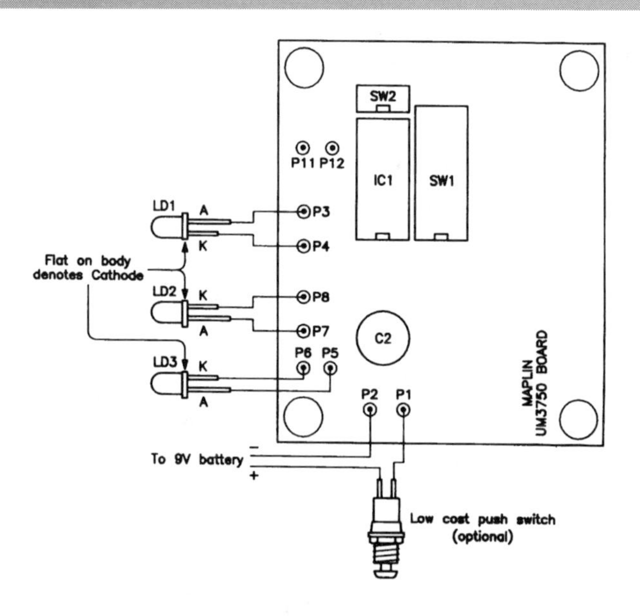

Figure 2.7 Transmitter PCB wiring

If the supply voltage is in excess of 11 volts, omit link LK1, insert RG1, ZD1 and R5 and change R3 and R4 to the values shown in Table 2.3. Two additional pins, P11(signal) and P12(ground) are available on the circuit board to allow the output of the UM3750 to be taken to the input of a fibre optic or RF link etc. These pins may also be used (without the UM3750 inserted) as the input for the infra-red link.

The circuit diagram for the receiver is given in Figure 2.8 and Figure 2.9 shows the PCB wiring. This circuit can be driven from a supply voltage between 6 volts and 35 volts. It should be noted that at supply voltage levels in excess of 11 volts, omit link LK1, and insert RG1, ZD1

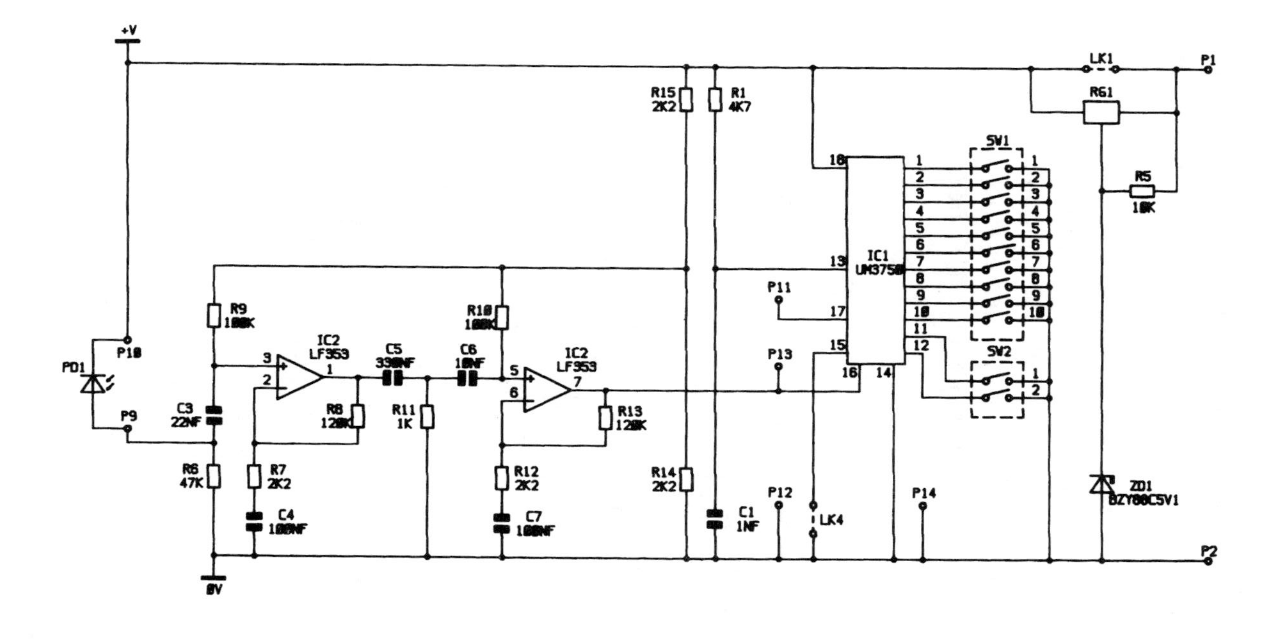

Figure 2.8 Receiver circuit diagram

and R5. The receiver output is taken from pins P11(signal) and P12(ground). This pin is normally held high at logic 1, but, on receipt of a valid code, changes to logic 0 until the valid code is removed, when it reverts back to logic 1. Pins P13(signal) and P14(ground) are provided, allowing the output from fibre optic/RF receivers etc. to be connected to the input of the UM3750 (do *not* insert IC2 or its associated components). These pins may also be used as the output of the infra-red link.

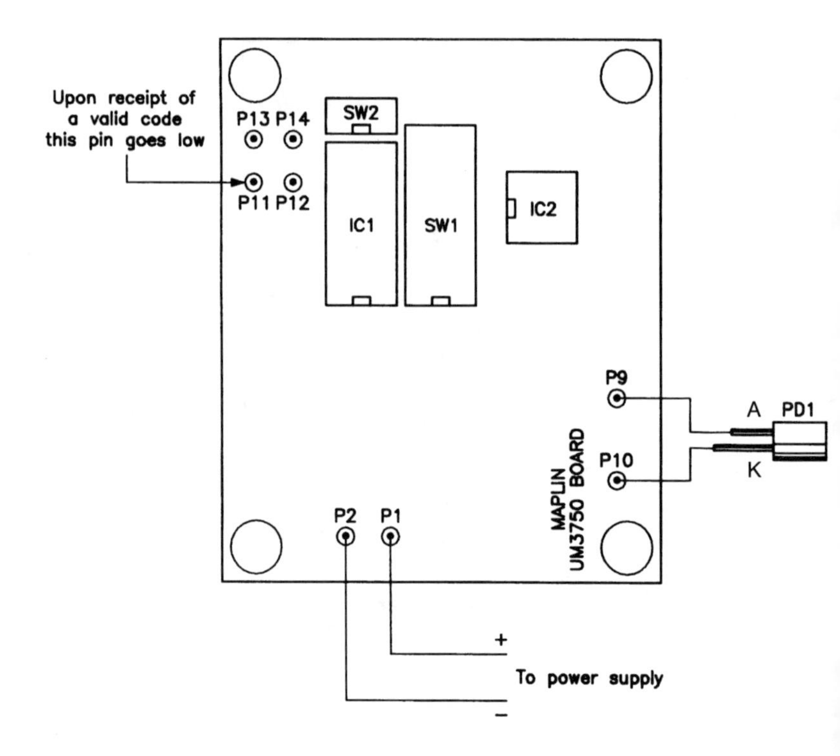

Figure 2.9 Receiver PCB wiring

Applications

A simple application for the device is an electric door lock. In order to open the door you aim the transmit unit at a box mounted on the wall. If your transmitted code matches the code set in the receiver then the door lock will operate. The typical wiring of such a system is shown in Figure 2.10. Note that as this system operates at 12

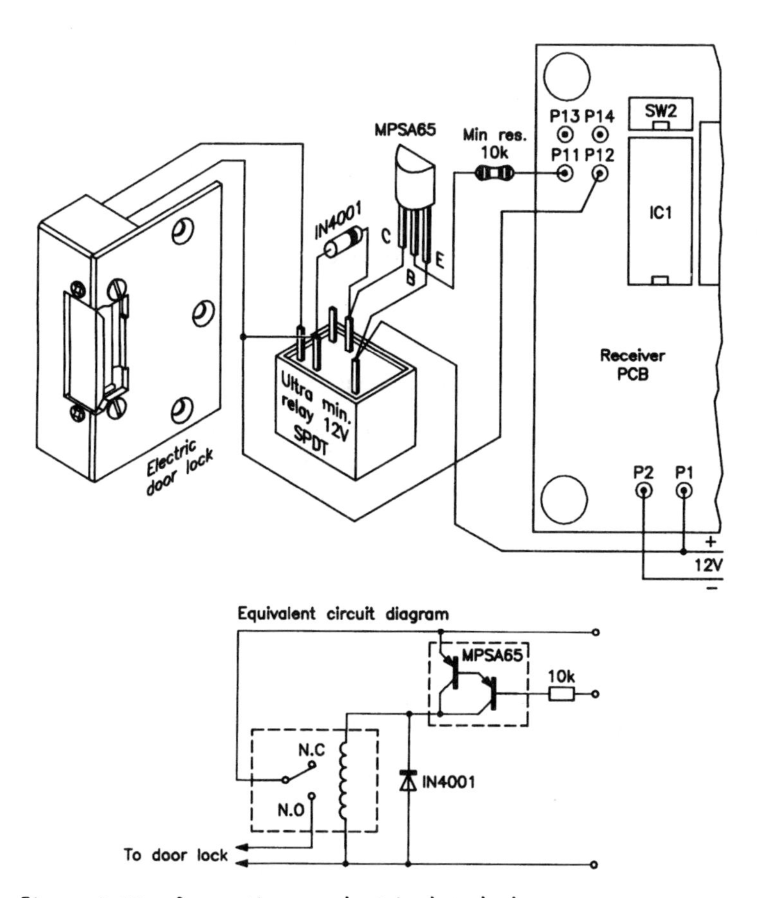

Figure 2.10 Connecting an electric door lock

volts, RG1, ZD1 and R5 need to be inserted. If mounting the receiver out-of-doors it is best to shield the photodiode from direct sunlight by mounting it recessed (as shown in Figure 2.11).

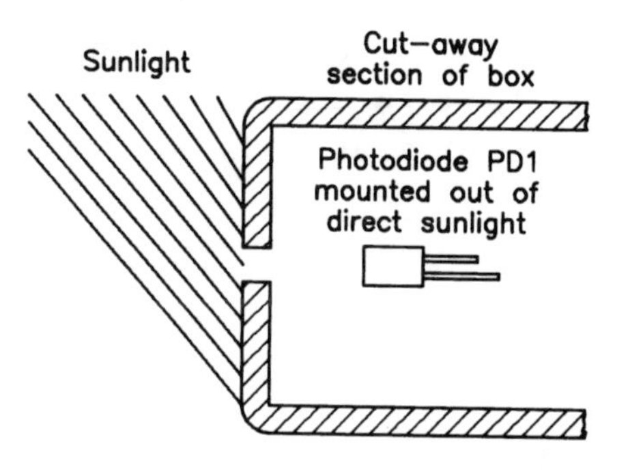

Figure 2.11 Mounting the photodiode PD1

A further application of the UM3750 would be to remotely activate/deactivate an alarm system. In an ideal system the user would press the *transmit* button once to activate the alarm and again to deactivate it. Most alarm systems require the active line to be logic 1 when the alarm is activated, and logic 0 when deactivated. The output of the UM3750 is, as previously stated, normally logic 1 and only changes to logic 0 for the duration of a valid code being received. Just connecting this output to the alarm system would result in it being deactivated for as long as somebody stood outside the premises with their finger firmly pressing the *transmit* button. The action of deactivating an alarm generally means that entry to the building is required, which can be quite difficult if you have to stand outside holding the transmit button! By placing a bistable circuit between the receiver and the alarm system, every depression of the *transmit* button will toggle the alarm between its activated and deactivated states. Figure 2.12 shows the basic configuration for such a system. Due to the vast number of alarm systems available it is beyond the scope of this chapter to describe how to connect the UM3750 to your alarm system.

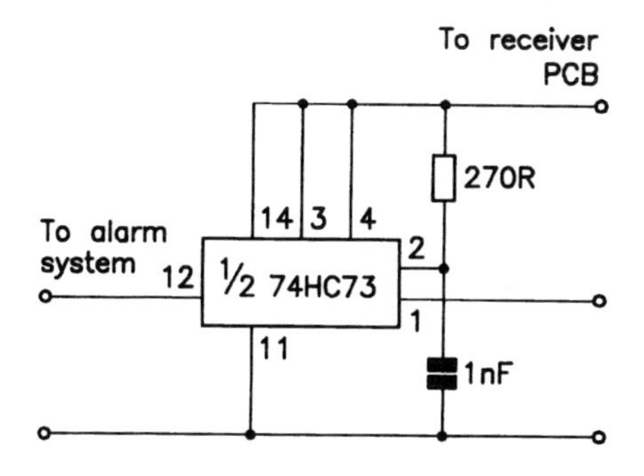

Figure 2.12 Connecting the UM3750 to an alarm system

Finally, Table 2.4 shows the specification for the prototype circuit built using the PCB.

Operating voltage	6 V–35 V
Operating frequency	400 kHz
Current consumption of transmitter	115 mA
Current consumption of receiver	7 mA
Effective range	40 cm–2 m

Table 2.4 Specification of prototype

UM3750 parts list

Transmitter

Resistors — All 1% 0.6 W metal film

R1,2	4k7	2	(M4K7)
R3	33 Ω(see text)	1	(M33R)
R4	18 Ω (see text)	1	(M18R)
R5	10 k (optional)	1	(M10K)

Capacitors

C1	1 nF monolithic	1	(RA39N)
C2	100 μF SMPS	1	(JL49D)

Semiconductors

IC1	UM3750	1	(UK77J)
TR1	BC337	1	(QB68Y)
RG1	μA78M05UC (optional)	1	(QL28F)
ZD1	BZY88C5V1 (optional)	1	(QH07H)
LD1	mini LED red	1	(WL32K)
LD2,3	infrared emitter	2	(YH70M)

Links

LK1	see text
LK2	fitted
LK3	fitted
LK4	not fitted

Miscellaneous

P1–8, 11,12	pins 2145	1	(FL24B)

SW1	10-way SPST DIL switch	1	(FV45Y)
SW2	SPST dual DIL switch	1	(XX26D)
	push switch	1	(FH59P)
	DIL socket	1	(HQ76H)
	18-pin PP3 battery clip	1	(HF28F)
	K9VHZ zinc chloride battery	1	(FK62S)
	PC board	1	(GE33L)
	constructors' guide	1	(XH79L)

Receiver

Resistors — All 1% 0.6 W metal film

R1	4k7	1	(M4K7)
R5	1 k (optional)	1	(M1K)
R6	47 k	1	(M47K)
R7,12, 14,15	2k2	4	(M2K2)
R8,13	120 k	2	(M120K)
R9,10	100 k	2	(M100K)
R11	1 k	1	(M1K)

Capacitors

C1	1 nF monolithic	1	(RA39N)
C3	22 nF monolithic	1	(RA45Y)
C4,7	100 nF monolithic	2	(RA49D)
C5	330 nF monolithic	1	(RA51F)
C6	10 nF monolithic	1	(RA44X)

Semiconductors

IC1	UM3750	1	(UK77J)
IC2	LF353	1	(WQ31J)
RG1	μA78M05UC (optional)	1	(QL28F)

ZD1	BZY88C5V1	1 (QH07H)
PD1	infra-red photodiode	1 (YH71N)

Links

LK1	see text
LK2	not fitted
LK3	not fitted
LK4	fitted

Miscellaneous

P1,2, 9–14	pins 2145	1 (FL24B)
SW1	SPST 10 way DIL switch	1 (FV45Y)
SW2	SPST dual DIL switch	1 (XX26D)
	18-pin DIL socket	1 (HQ76H)
	8-pin DIL socket	1 (BL17T)
	PC board	1 (GE33L)
	constructors' guide	1 (XH79L)

Optional

10 k min res	1 (M10K)
MPSA65	1 (QH61R)
1N4001	1 (QL73Q)
electric door lock	1 (YU89W)
12 V SPDT ultra-min relay	1 (YX94C)
270 Ω min res	1 (M270R)
1 nF monolithic	1 (RA39N)
74HC73	1 (UB18U)

ZN419/409 precision servo

The ZN419 precision servo IC (also supplied as the ZN409) is specifically designed for use with pulse width operated servo mechanisms, used in a variety of control applications. A low external component count and relatively low power consumption make the device ideal for use in model boats, aircraft and cars where battery life, space and weight are of prime consideration.

In addition to the role of a servo driver, it is also possible to use the IC in motor speed control applications. The device is supplied in a 14-pin DIL package. Figure 2.13 shows the IC pin-out and Table 2.5 shows some typical electrical characteristics for the device.

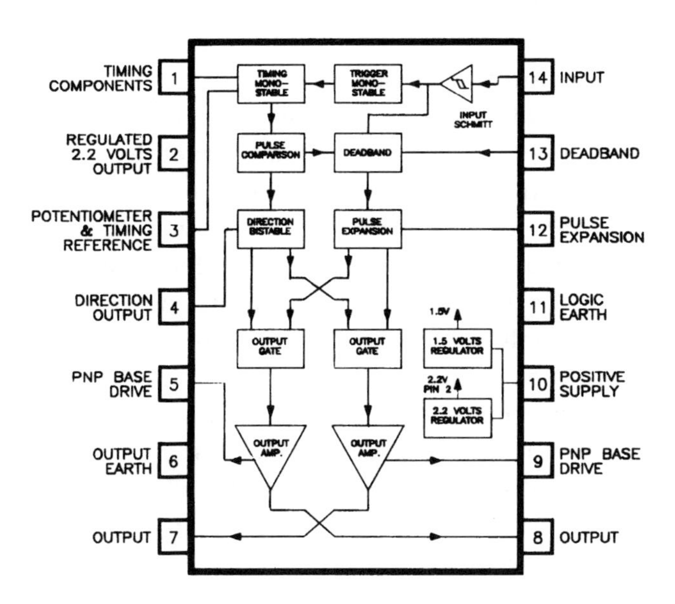

Figure 2.13 IC pin-out

Parameter	*Conditions*	*Min*	*Typ*	*Max*
Supply voltage		3.5 V	5 V	6.5 V
Supply current	Quiescent	4.6 mA	6.7 mA	10 mA
Input resistance		20 k	27 k	35 k
Input current		350 μA	500 μA	650 μA
Lower input threshold	(Pin 14)	1.15 V	1.25 V	1.35 V
Upper input threshold	(Pin 14)	1.4 V	1.5 V	1.6 V
Minimum output pulse		2.5 ms	3.5 ms	4.5 ms
Output saturation voltage	Load current = 400 mA		300 mV	400 mV

Table 2.5 ZN419 typical electrical characteristics

General description

A typical control system is that based on the operation of a joy-stick to vary the pulse width of a timing circuit. Large numbers of pulses are multiplexed by time division and are used to modulate a radio control transmitter. At the receiver, the received signal is demodulated and a train of pulses is produced to control a servo. The servo consists of a motor driven reduction gearbox which has a potentiometer coupled to the output shaft. The servo potentiometer produces an output which corresponds to the position of the output shaft. The output from the potentiometer is used to control the pulse width of a timing monostable which forms part of the ZN419 IC. An internal pulse comparison stage compares the input pulse width with that of the monostable pulse and produces an output which determines the correct phase of the on-chip power amplifier. Another output from the pulse comparison circuit drives a pulse expansion circuit which expands the difference between the input and monostable pulses. The difference pulse is used to drive the motor in such a direction as to reduce the difference so that the servo takes up a position which corresponds to that of the controller joy-stick.

In addition to driving servo motors, the ZN419 can also be used for motor speed control. In this application the IC acts as a linear pulse width amplifier. The motor is driven with a train of pulses which have a variable pulse width such that the speed of the motor can be controlled between zero and maximum. The ZN419 uses fixed timing components and a fixed resistor is used in place of the servo potentiometer. The centre of the range monostable period represents zero motor speed, and pulses less than or greater than this drive the motor in the forward or reverse direction respectively. A direction output is produced on pin 4 of the IC and this may be used to control a relay to determine the motor direction. The pulse expansion components should be chosen to provide a suitable relation between the controller joy-stick position and the speed of the motor.

A *deadband* is required around the centre of this range, between the minimum forward and reverse positions, where no power is applied to the motor. The IC contains an internal deadband control circuit and the size of the deadband is adjustable to suit different requirements.

Because of the high current consumption of many motors, using the device in motor speed control applications usually requires a separate high current power supply to increase the operating time between recharging the batteries.

IC power supply

The IC requires a 4 V to 6.5 V (absolute maximum) single rail power supply which is capable of supplying a current of up to 10 mA. As the device is primarily designed for use in radio controlled models the supply is usually derived from a battery. For reliable operation it is recommended that high frequency supply decoupling is used close to the IC to prevent high frequency voltage spikes on the supply rail.

PCB available

A high quality, double sided, fibreglass PCB with screen printed legend is available as an aid to constructors wishing to use the ZN419 IC. The PCB may be utilised in the construction of either a servo driver or a motor speed control circuit. Figure 2.14 shows the combined circuit diagrams for both the speed controller and serve driver options. The PCB legend is shown in Figure 2.15.

In order to make the PCB as small as possible, components are mounted on both sides of the board. The PCB is marked *side A* and *side B* for ease of identification. When constructing either circuit, the components mounted on side A should be fitted first, followed by the components on side B. It is necessary to solder some components on the same side of the PCB as they are mounted, instead of the normal *other side*. This is particularly the case with R5, R8, R11, R13, R14, R15, R16 and D4. Protruding component leads should be cut as close to the PCB as possible so that they do not obstruct components on the other side of the board.

Servo driver option

Figure 2.16 shows the circuit diagram of the servo driver application. If building this circuit, reference should be made to the servo driver parts list only and *not* to the motor speed controller parts list which shows a different set of component values. Figure 2.17 shows the wiring diagram for the module when used as a servo driver.

The circuit is primarily designed to operate from a 6 V battery but may be powered from any supply voltage between 4 V and 6 V. Power supply connections are made to P1(+V) and P2(0 V). A pulse width modulated input is required to drive the module, with a variable pulse width between 0.2 ms and 2.5 ms as illustrated in Figure 2.18 and Figure 2.19. A typical test circuit is shown in Figure 2.20. Please note: LK1 and LK2 are *not* fitted for the servo driver application, but *both* LK3 and LK4 should be fitted.

Motor speed control option

Figure 2.21 shows the diagram of the motor speed control circuit. When building this circuit reference should be made to the motor speed control parts list only and *not* to the servo driver parts list which shows a different set of component values. Figure 2.22 shows the wiring diagram for the module when used as a motor speed control circuit.

The circuit has two separate power supply rails. A 4 to 6 V power supply is used for IC1 and associated compo-

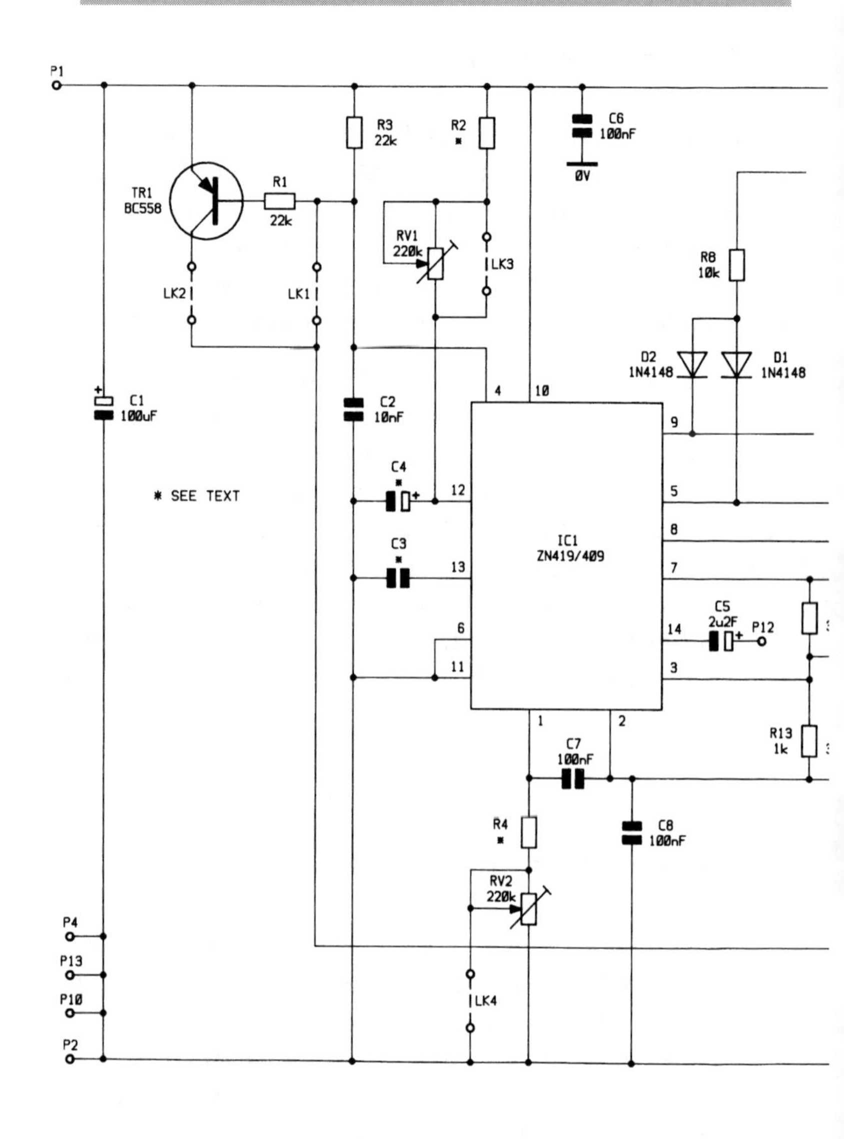

Figure 2.14 Combined circuit diagram of module

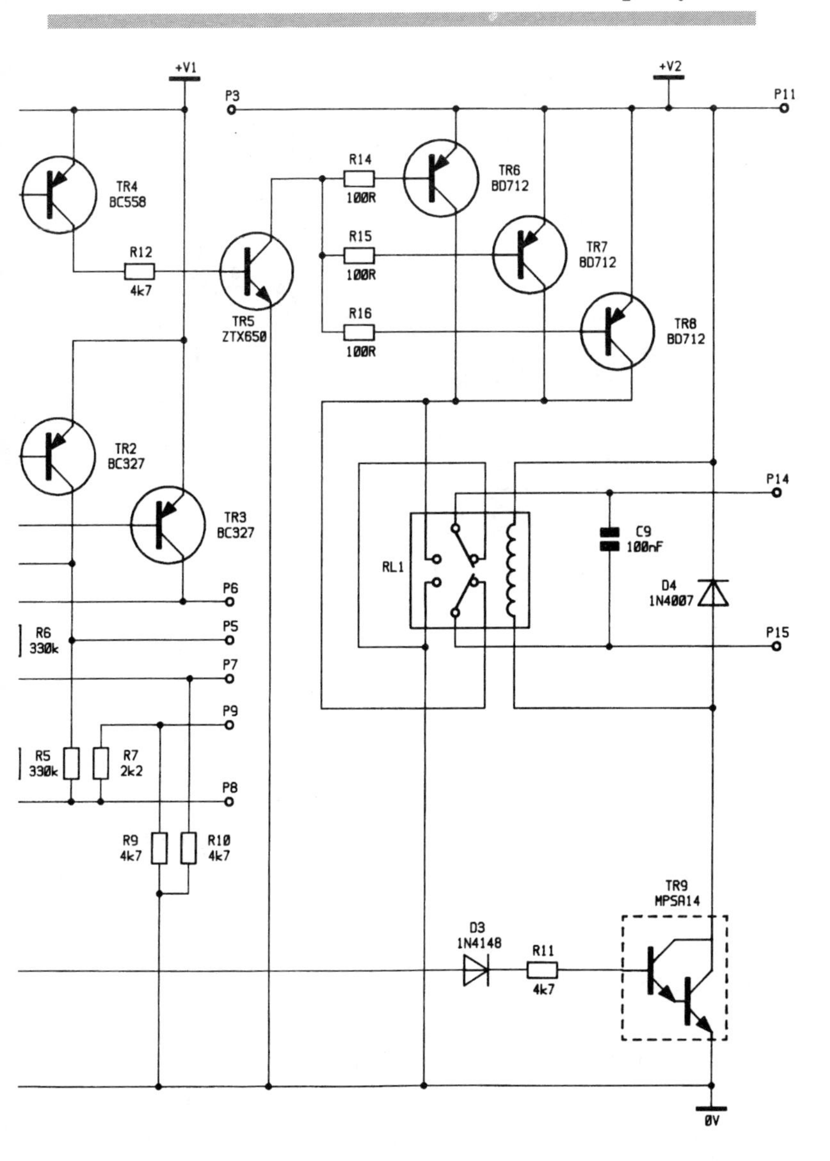

Figure 2.14 (Continued)

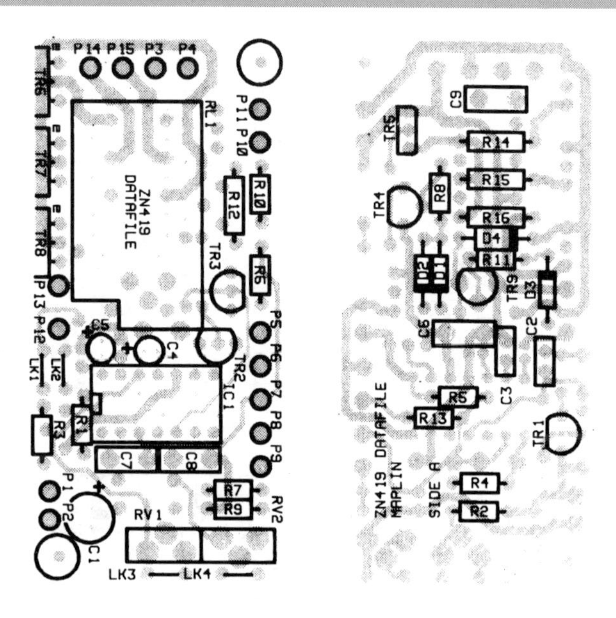

Figure 2.15 Legend on both sides of PCB

nents, and the connections from this supply are made to P1(+V) and P2(0 V). An additional supply (usually provided by a rechargeable Ni-Cd pack) is required for the motor drive circuitry and connections from this supply are made to P3(+V) and P4(0 V). The supply voltage for the motor may be between approximately 6 V and 8 V at continuous currents up to 5 A using the components specified; however, higher voltages and currents can be accommodated using different component values. In particular, the power handling capability of relay RL1 and resistors R14–R16 should be taken into consideration as well as the power dissipation of transistors TR5–TR8. A suitable 12 V relay to use in place of RL1 is FJ43W.

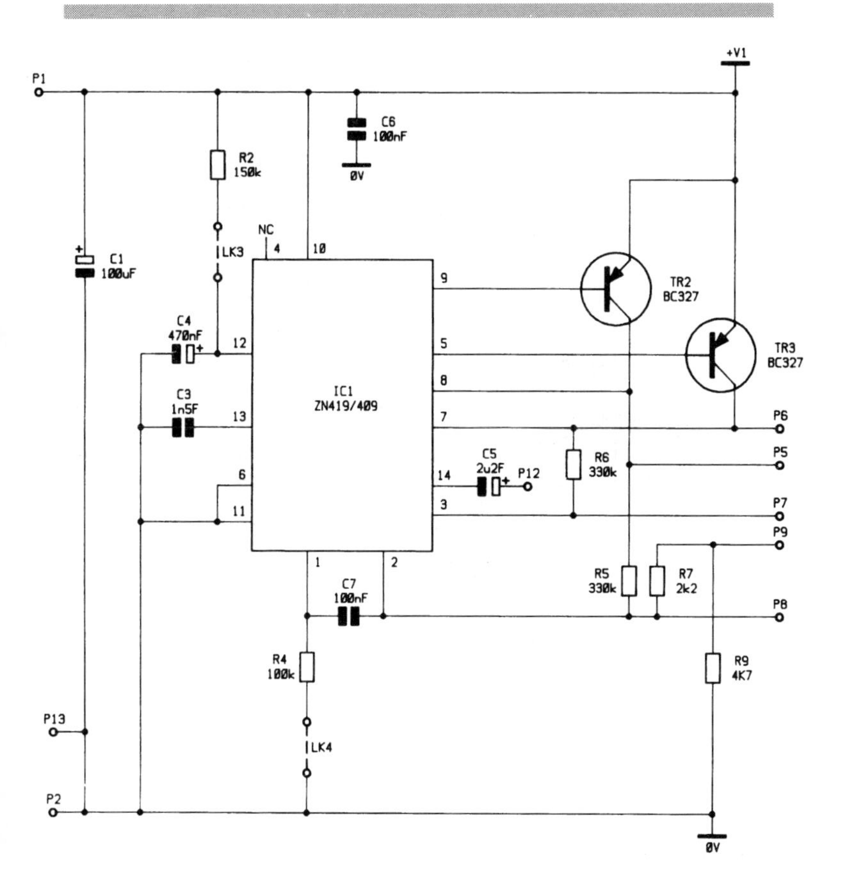

Figure 2.16 Circuit diagram of servo application

Power transistors TR6–TR8 are positioned close to the edge of the PCB), to allow the installation of heatsinks as required. There is a small *e* on the legend denoting the emitter of the transistor for reference purposes. If and to what extent a heatsink is required is really determined by the power consumption of the motor being

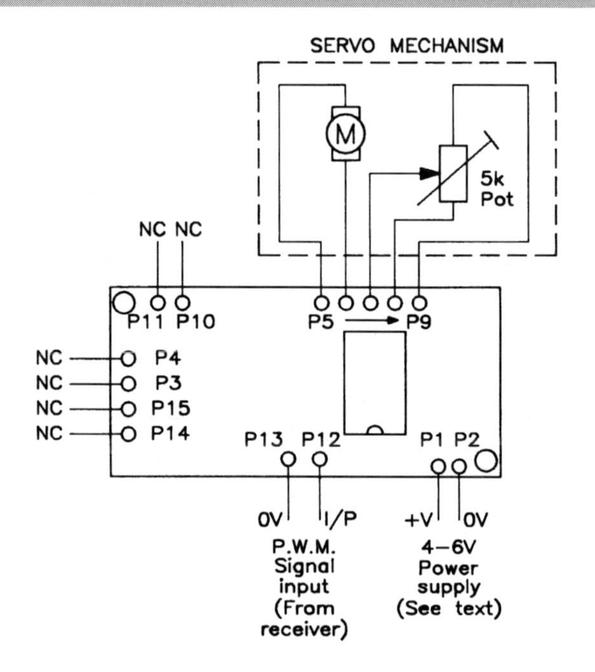

Figure 2.17 Servo driver wiring diagram

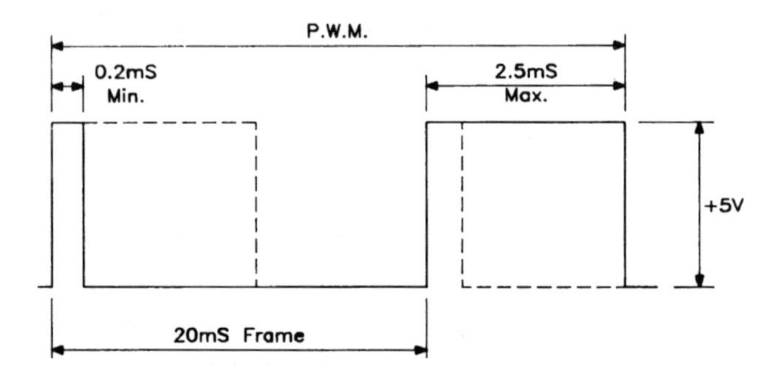

Figure 2.18 Control signal

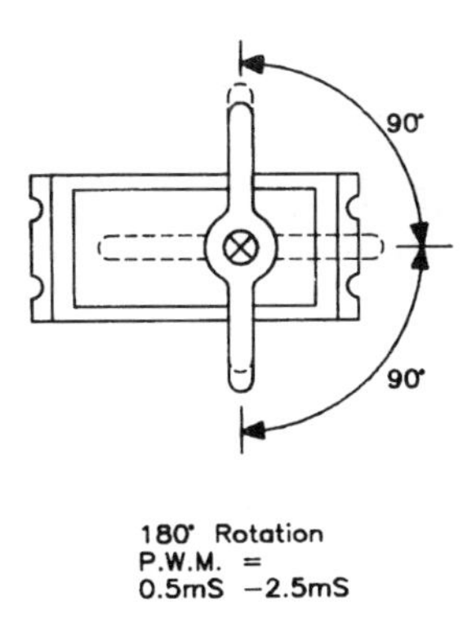

Figure 2.19 Servo operation

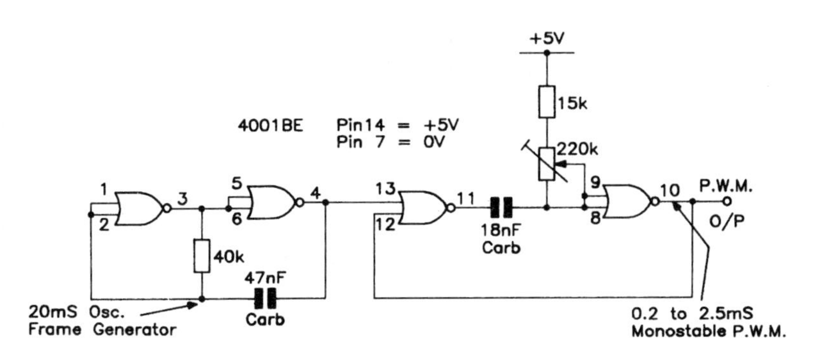

Figure 2.20 Typical test circuit

driven; higher power motors will obviously require additional heatsinking. If the model has a metal chassis, it is often possible to use this as a heatsink for the motor drive transistors. It should be remembered though that the transistor heatsink tags are at collector potential and if necessary should be insulated from the heatsink using a suitable bush and insulating washer such as stock code WR23A.

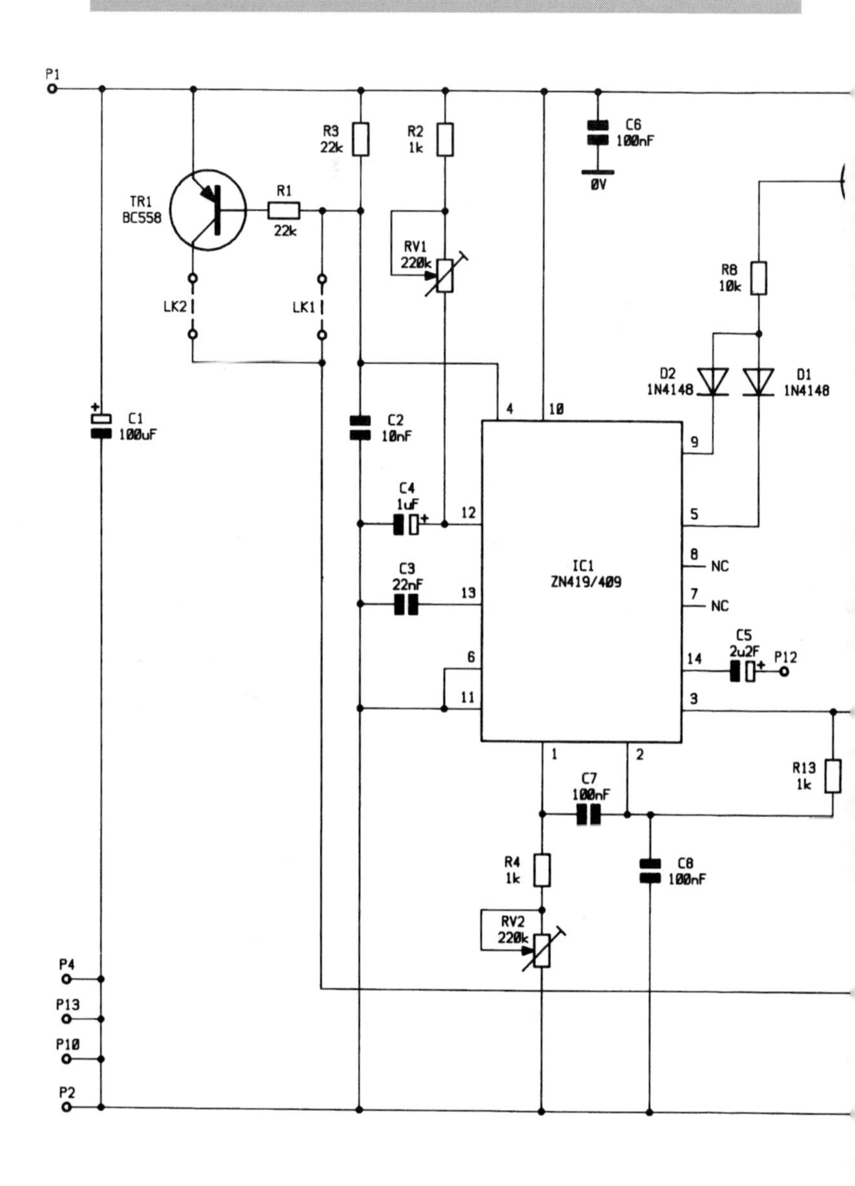

Figure 2.21 Circuit diagram of motor speed control application

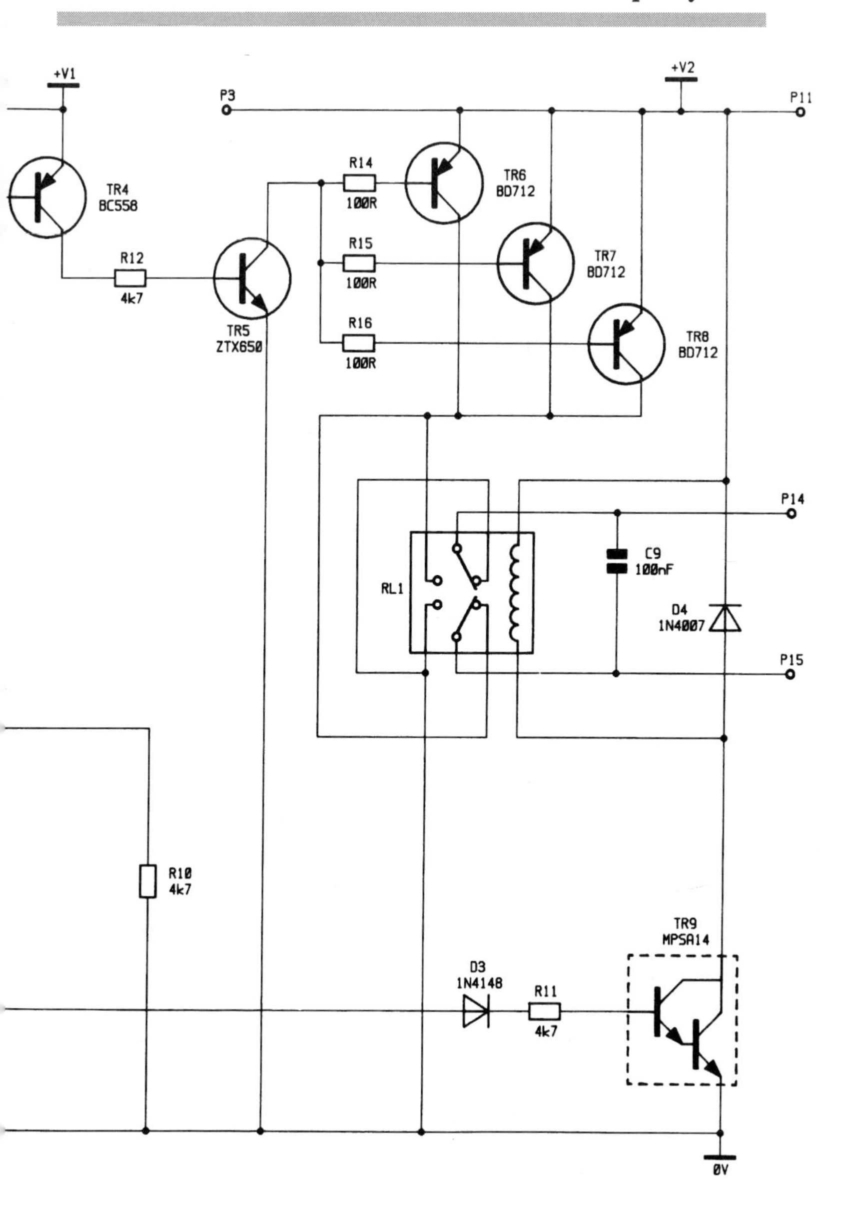

Figure 2.21 (Continued)

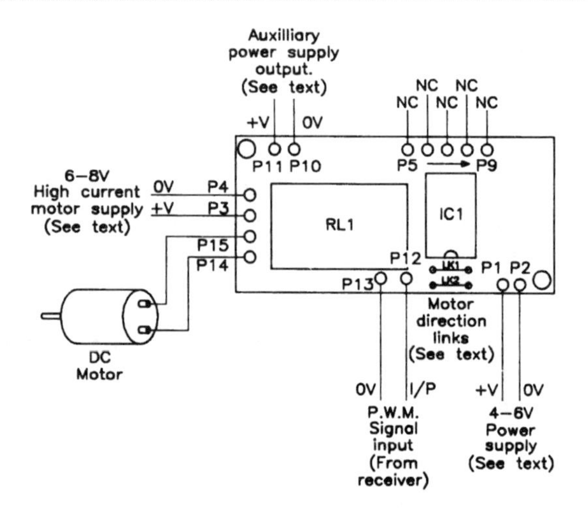

Figure 2.22 Motor speed control wiring diagram

The direction of rotation of the motor is set by fitting either wire link LK1 or LK2. Please note LK1 and LK2 *should never be used together at the same time!* LK3 and LK4 are not used in the motor speed control application.

Two preset resistors are used to align the module. The relationship between the control stick and the speed of the motor is set using RV1. Because both forward and reverse motor drive are required, a *no drive* or zero position is needed; this is determined by adjusting RV2. Some experimentation is necessary to optimise the parameters of the module for individual applications.

Finally, Table 2.6 shows the specifications for both the servo driver and speed control options from the prototype module.

Servo driver application

Power supply voltage	4 V–6 V
Power supply current (quiescent)	8 mA at 6 V
Input pulse width	0.2 ms–2.5 ms

Motor speed control application

Parameter	Condition	Value
Power supply voltage	Low current supply (P1 and P2)	4 V–6 V
	High current supply (P3 and P4)	6 V–8 V
Power supply current	P1 and P2 (quiescent)	8 mA at 6 V
	P3 and P4	See text
Input pulse width		0.2 ms–2.5 ms

Table 2.6 Specification of prototype

Servo driver parts list

Resistors — All $^1/_8$ W 5% carbon film

R1,3	not fitted		
R2	150 k micro res	1	(U150K)
R4	100 k micro res	1	(U100K)
R5,6	330 k micro res	2	(U330K)
R7	2k2 micro res	1	(U2K2)
R8	not fitted		
R9	4k7 micro res	1	(U4K7)
R10,11	not fitted		
R12	not fitted		
R13	not fitted		
R14,15, 16	not fitted		
RV1,2	not fitted		

Capacitors

C1	100 μF 10 V minelect	1	(RK50E)
C2	not fitted		
C3	1500 pF ceramic	1	(WX70M)
C4	470 nF 63 V minelect	1	(YY30H)
C5	2200 nF 63 V minelect	1	(YY32K)
C6,7	100 nF 16 V minidisc	2	(YR75S)
C8,9	not fitted		

Semiconductors

IC1	ZN419/409CE	1	(YH92A)
TR1,4	not fitted		

TR2,3	BC327	2(QB66W)
TR5	not fitted	
TR6,7,8	not fitted	
TR9	not fitted	
D1,2,3	not fitted	
D4	not fitted	

Miscellaneous

RL1	not fitted	
	14-pin DIL socket	1 (BL18U)
P1,2,5,6, 7,8, 9, 12,13	pin 2145	1 (FL24B)
LK1	not fitted	
LK2	not fitted	
LK3	wire link	fit wire link
LK4	wire link	fit wire link
	PC board	1 (GE83E)
	leaflet	1 (XK49D)
	constructors' guide	1 (XH79L)

Motor driver parts list

Resistors — All $^1/_8$ W 5% carbon film (unless specified)

R1,3	22 k micro res	2	(U22K)
R2,4,13	1 k micro res	3	(U1K)
R5,6	not fitted		
R7	not fitted		
R8	10 k micro res	1	(U10K)
R9	not fitted		
R10,11	4k7 micro res	2	(U4K7)
R12	4k7 (0.6 W 1% metal film)	1	(M4K7)
R14,15, 16	100 Ω (0.6 W 1% metal film)	3	(M100R)
RV1,2	220 k vert encl preset	2	(UH20W)

Capacitors

C1	100 μF 10 V minelect	1	(RK50E)
C2	10,000 pF ceramic	1	(WX77J)
C3	22,000 pF ceramic	1	(WX78K)
C4	1 μF 63 V minelect	1	(YY31J)
C5	2200 nF 63 V minelect	1	(YY32K)
C6,7,8,9	100 nF 16 V minidisc	4	(YR75S)

Semiconductors

IC1	ZN419/409CE	1	(YH92A)
TR1,4	BC558	2	(QQ17T)
TR2,3	not fitted		
TR5	ZTX650	1	(UH46A)

TR6,7,8	BD712	3 (WH16S)
TR9	MPSA14	1 (QH60Q)
D1,2,3	1N4148	3 (QL80B)
D4	1N4007	1 (QL79L)

Miscellaneous

RL1	min 6 V 6 A relay	1 (FJ42V)
	14-pin DIL socket	1 (BL18U)
P1,2,3,4, 10,11,12, 13,14,15	pin 2148	1 (FL24B)
LK1	wire link (see text)	
LK2	wire link (see text)	
LK3	not fitted	
LK4	not fitted	
	PC board	1 (GE83E)
	leaflet	1 (XK49D)
	constructors' guide	1 (XH79L)

LM331 voltage to frequency/ frequency to voltage converter

The LM331 is a simple voltage-to-frequency converter suitable for use in analogue-to-digital conversion, precision voltage-to-frequency conversion and many other applications. Figure 2.23 shows the IC pinout, and Table 2.7 shows typical characteristics for the device. When the IC is used as a voltage-to-frequency converter, it produces a pulse train which is linearly proportional to the applied input voltage. The device uses a temperature-compensated band-gap reference circuit to provide very good accuracy over the entire operating temperature range. Although the precision timer circuit has low bias currents, the response is sufficiently fast for 100 kHz voltage-to-frequency conversion (low bias currents often result in reduced switching speeds and restricted operating frequencies). The output of the device is capable of driving loads of between 5 V (i.e. TTL level) and 40 V, depending on the supply voltage, and is fully protected against short circuits to V_{CC}.

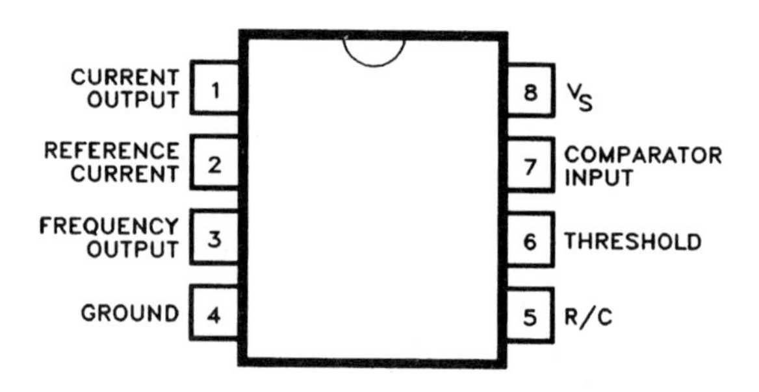

Figure 2.23 IC pin-out

Parameter	Conditions	Min	Typ	Max
Supply voltage	Absolute maximum			40 V
Supply current	5 V supply	1.5 mA	3 mA	6 mA
	40 V supply	2 mA	4 mA	8 mA
Operating ambient temperature range	Absolute maximum limits	0°C		70°C
Reference voltage		1.7 V	1.89 V	2.08 V
Change of gain with supply voltage	Supply voltage = 4.5 V to 10 V		0.01%V	0.1%V
	Supply voltage = 10 V to 40 V		0.006%V	0.06%V
Rated full scale frequency	Input voltage = 10 V	10 kHz		
Output current (pin 1)	Resistance pin 2 (Rs) = 14 kΩ; pin 1 voltage = 0 V	116 μA	136 μA	156 μA
Operating range of current		10 μA	to	500 μA

Table 2.7 Typical characteristics of LM331 voltage-to-frequency converter

Voltage-to-frequency conversion

Figure 2.24 shows the block diagram of a simple stand-alone voltage-to-frequency converter. Figure 2.25 shows a practical implementation of this. Resistor R_{in} has been included so that the bias current at pin 7 (80 nA typical) will cancel the effect of the bias current at pin 6, helping to provide minimum frequency offset.

The resistance at pin 2 (R_s) is made up of a 12 k fixed resistor and a 5 k pre-set resistor, which is used to trim the gain tolerance of the IC and associated components;

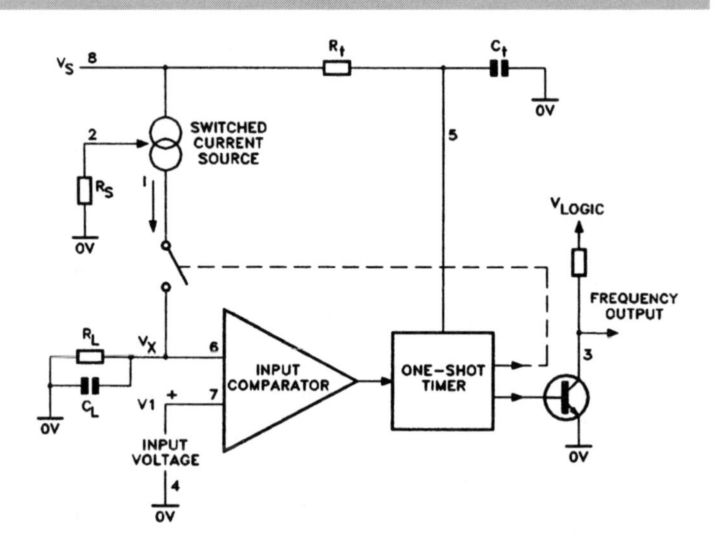

Figure 2.24 Block diagram of a simple voltage-to-frequency converter

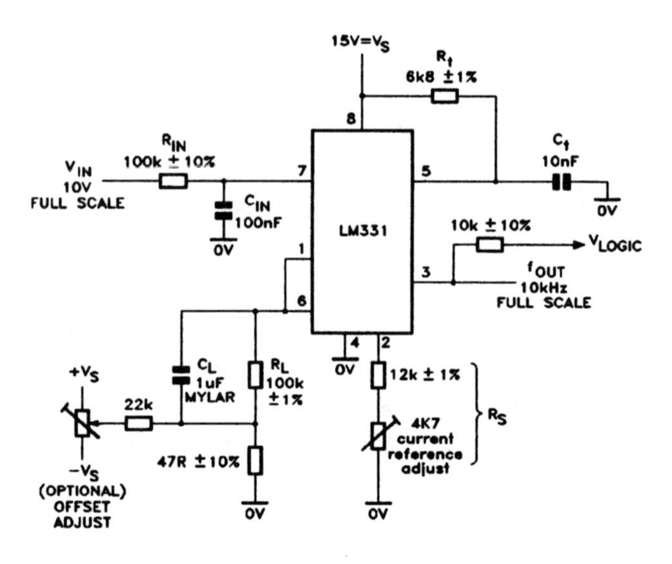

Figure 2.25 Example of a practical voltage-to-frequency converter

it is advantageous to use close tolerance resistors in this application. The capacitors used should feature a low dielectric absorption, depending on the desired temperature characteristics; polystyrene and polypropylene types are suitable. Capacitor C_{in} is connected between pin 7 and ground to act as a simple input filter; a value of between 10 nF and 1 μF can be used for this purpose. When the time constants of the RC networks on pins 6 and 7 are matched, a given voltage step at V_{in} will result in a corresponding step in output frequency. If C_{in} is much less than C_L, a change in voltage at V_{in} may cause the output (F_{out}) to stop momentarily. A 47 Ω resistor is connected in series with C_L to produce a hysteresis effect; this helps to improve the overall linearity.

Frequency-to-voltage conversion

The LM331 may also be used to provide a simple frequency-to-voltage converter, and a typical example of this application is shown in Figure 2.26. In this application, a pulse at the input (F_{in}) is differentiated by an RC network and the negative-going edge at pin 6 causes the input comparator to trigger the timing circuit. As with the voltage-to-frequency converter, the average current flowing from pin 1 is:

$$I \times (1.1\, R_t C_t) \times f$$

where I is the current flowing from pin 2 and f is the input frequency. I can be calculated by the following formula:

$$I = V_{ref} / R_s$$

where V_{ref} is the reference voltage at pin 2 (typically 1.89 V) and Rs is the resistance between pin 2 and 0 V. The current is filtered by a simple RC network. The ripple is typically less than 10 mV peak, but the response of the circuit is inherently slow with a 0.1 second time constant and a settling time of 0.7 seconds (to 0.1% accuracy).

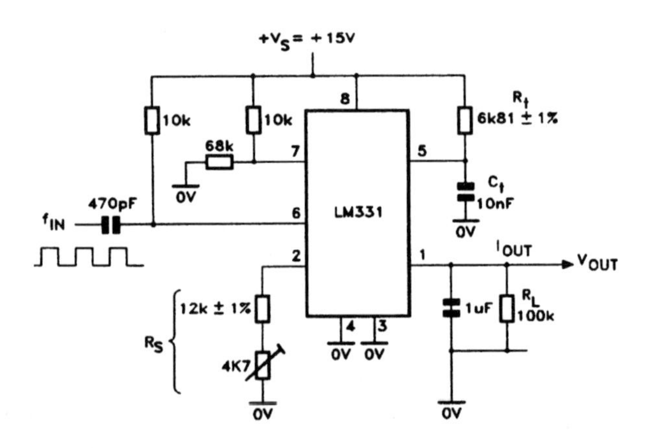

Figure 2.26 Example of a simple frequency-to-voltage converter

Kit available

A kit of parts is available, which covers the two basic applications of the LM331 mentioned in this chapter.

Voltage-to-frequency converter

The kit contains the components necessary to build a simple voltage-to-frequency converter; see *Voltage-to-Frequency Converter Parts List*. Figure 2.27 shows the circuit diagram of the module, while Figure 2.29 shows the legend. Additional parts are included in the kit to allow the module to be configured as a frequency-to-voltage converter, and in this case it is necessary to fit a different selection of components.

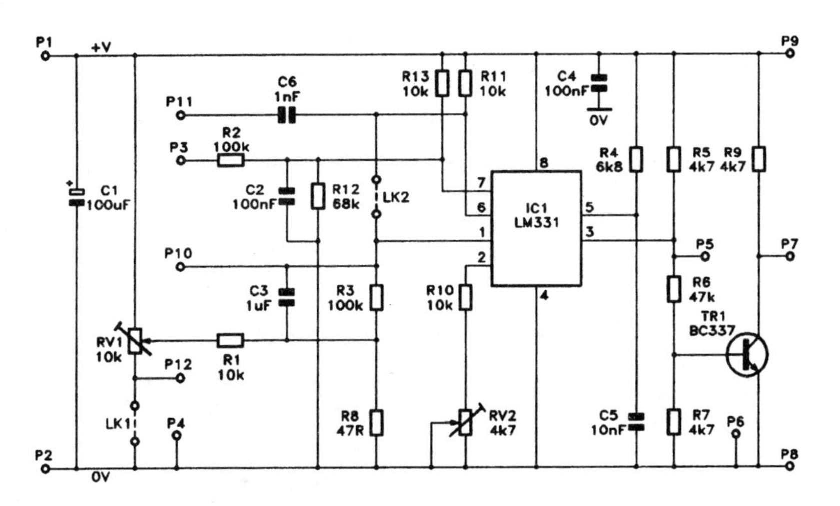

Figure 2.27 Circuit diagram of module

Figure 2.28 shows the circuit diagram of the module configured as a voltage-to-frequency converter. An input voltage is applied between P3(input) and P4(0 V); the corresponding output frequency may be taken between P5(normal output) and P6(0 V), or P7(inverted

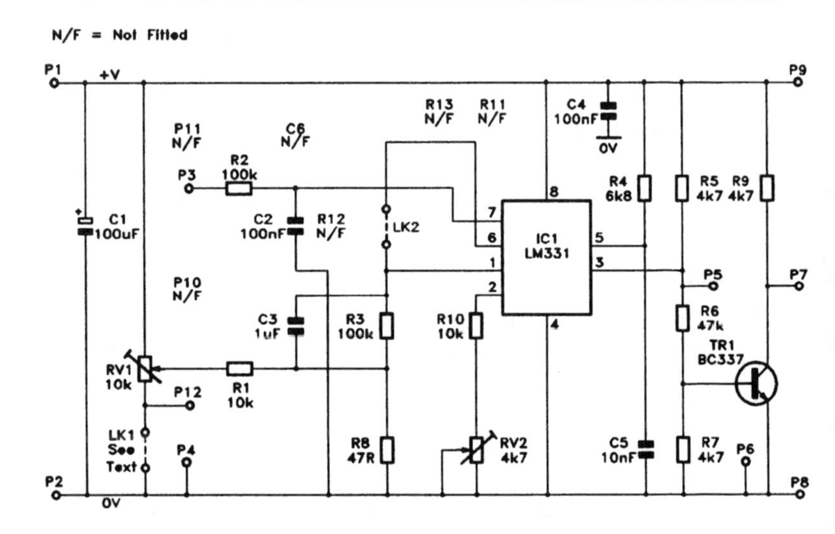

Figure 2.28 Circuit diagram of a voltage-to-frequency converter module

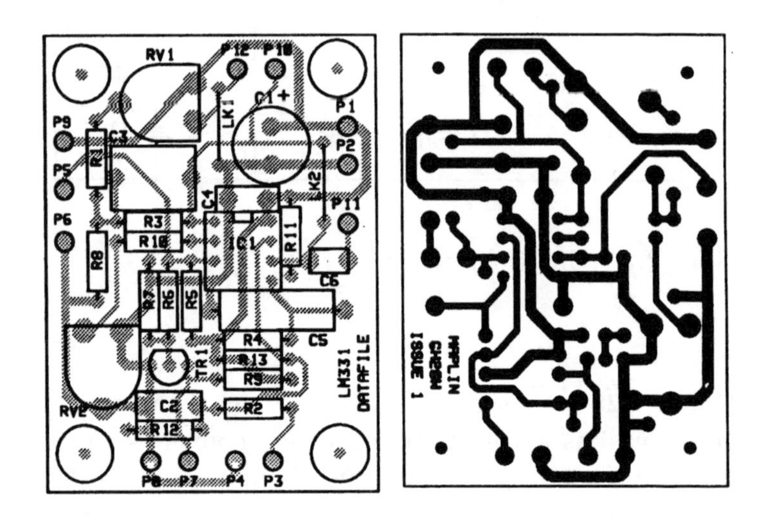

Figure 2.29 PCB legend and track

output) and P8(0 V). Please note: the normal output at P5 does not swing all the way to +V in this circuit configuration. There are additional +V and 0 V pins on the PCB for increased versatility. Wiring information for the voltage-to-frequency converter application is shown in Figure 2.30.

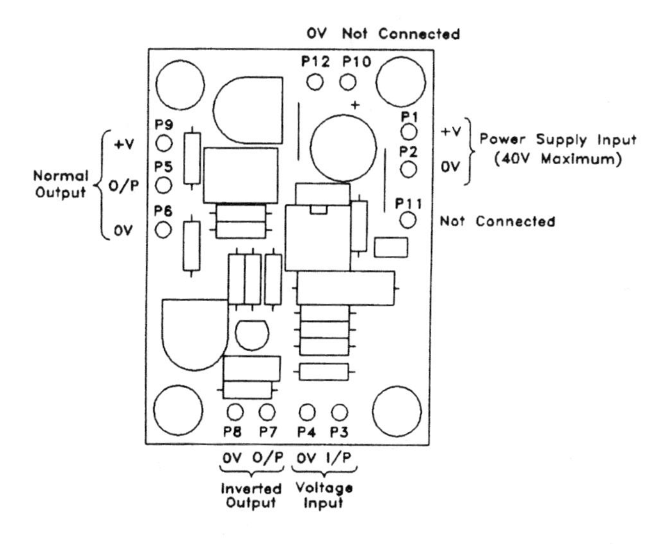

Figure 2.30 Wiring diagram for voltage-to-frequency converter module

Link 1 is normally fitted; however, it can be omitted if RV1, the conversion-offset fine-adjustment pre-set, is to be referenced to an external negative voltage (–10 V max.), for increased adjustment range. This external voltage is applied to pin P12. RV2 sets the reference current.

Frequency-to-voltage converter

Figure 2.31 shows the circuit diagram of the module configured as a frequency-to-voltage converter. The *Frequency-to-Voltage Converter Parts List* identifies the components required for assembly. In this application, an input frequency (1 Hz to 10 kHz) is applied between P11(input) and 0 V, and an output voltage is taken between P10 and 0 V. The output voltage is directly related to the current flowing through R3, the value of which could be changed, if necessary, to provide different output level swings. Because of the high impedance nature of the output, it will normally be necessary to provide additional buffering — unless the circuit is driving a load

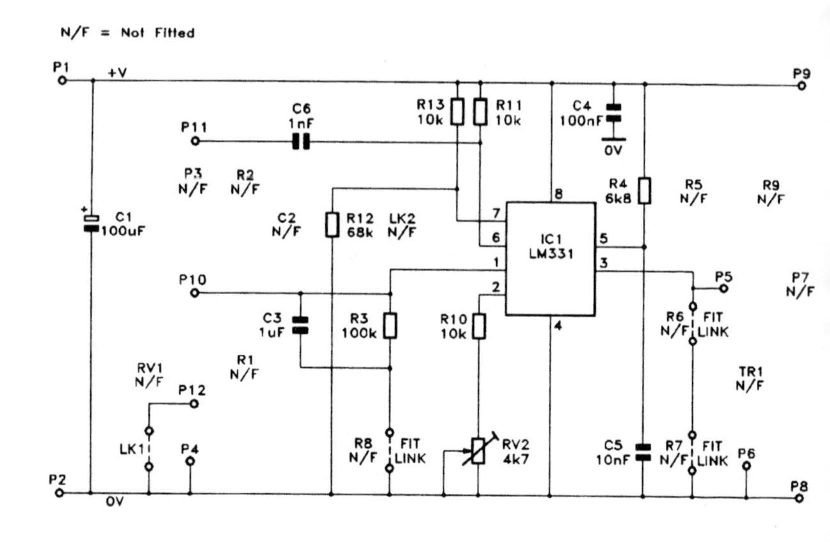

Figure 2.31 Circuit diagram of a frequency-to-voltage converter module

with a considerably higher impedance than R3. If the circuit is used to drive lower impedances, the output level will be considerably reduced. Figure 2.32 shows wiring information for the frequency-to-voltage converter application.

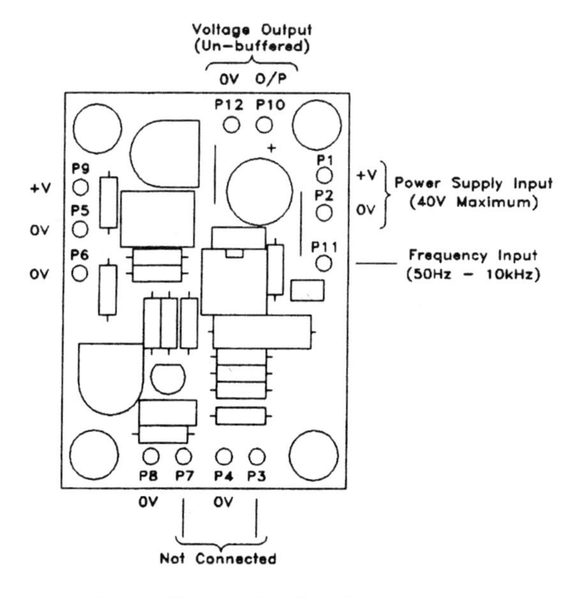

Figure 2.32 Wiring diagram for frequency-to-voltage converter module

Tables 2.8 and 2.9 show the specification of each of the prototype circuits.

Please note: Only the parts itemised in the parts list of the specific application that you are building should be fitted, i.e. if you are building the frequency-to-voltage converter, only fit the parts shown in the frequency-to-voltage converter parts list and so on.

Power supply voltage	5 V to 40 V
Power supply current (at 12 V)	8 mA
Output frequency range (at 12 V)	50 Hz to 10 kHz
Input voltage range (at 12 V)	0.05 V to 10 V

Table 2.8 Specification of prototype voltage-to-frequency converter

Power supply voltage	5 V to 40 V
Power supply current (at 12 V)	6 mA
Output voltage range (at 12 V)	0.05 V to 10 V
Input frequency range (at 12 V)	50 Hz to 10 kHz
Input level	As supply voltage

Table 2.9 Specification of prototype frequency-to-voltage converter

Voltage-to-frequency converter parts list

Resistors — All 0.6 W 1% metal film (unless specified)

R1,10	10 k	2 (M10K)
R2,3	100 k	2 (M100K)
R4	6k8	1 (M6K8)
R5,7,9	4k7	3 (M4K7)
R6	47 k	1 (M47K)
R8	47 Ω	1 (M47R)
R11,12, 13	not fitted	
RV1	10 k hor encl preset	1 (UH03D)
RV2	4k7 hor encl preset	1 (UH02C)

Capacitors

C1	100 µF 63 V PC elect	1 (FF12N)
C2	100 nF polylayer	1(WW41U)
C3	1 µF polylayer	1(WW53H)
C4	100 nF disc ceramic	1 (BX03D)
C5	10 nF 1% polystyrene	1 (BX86T)
C6	not fitted	

Semiconductors

TR1	BC337	1 (QB68Y)
IC1	LM331	1 (UL47B)

Miscellaneous

	8-pin DIL socket	1 (BL17T)
	pin 2145	1 (FL24B)

	PCB	1 (GH20W)
	instruction leaflet	1 (XT84F)
	constructors' guide	1 (XH79L)
LK1	fitted (see text)	
LK2	fitted	

Frequency-to-voltage converter parts list

Resistors — All 0.6 W 1% metal film (unless specified)

R1,2,5,9	not fitted	
R3	100 k	1 (M100K)
R4	6k8	1 (M6K8)
R6,7,8	fit link	
R10,11, 13	10 k	3 (M10K)
R12	68 k	1 (M68K)
RV1	not fitted	
RV2	4k7 hor encl preset	1 (UH02C)

Capacitors

C1	100 μF 63 V PC elect	1 (FF12N)
C2	not fitted	
C3	1 μF polylayer	1 (WW53H)
C4	100 nF disc ceramic	1 (BX03D)
C5	10 nF 1% polystyrene	1 (BX86T)
C6	1 nF ceramic	1 (WX68Y)

Semiconductors

TR1	not fitted	
IC1	LM331	(UL47B)

Miscellaneous

	8-pin DIL socket	1 (BL17T)
	pin 2145	1 (FL24B)
	PCB	1 (GH20W)
	instruction leaflet	1 (XT84F)
	constructors' guide	1 (XH79L)
LK1	fitted	
LK2	not fitted	

LM331 data file parts list

Resistors — All 0.6 W 1% metal film (unless specified)

R1,10, 11,13	10 k	4 (M10K)
R2,3	100 k	2 (M100K)
R4	6k8	1 (M6K8)
R5,7,9	4k7	3 (M4K7)
R6	47 k	1 (M47K)
R8	47 Ω	1 (M47R)
R12	68 k	1 (M68K)
RV1	10 k hor encl preset	1 (UH03D)
RV2	4k7 hor encl preset	1 (UH02C)

Capacitors

C1	100 μF 63 V PC elect	1 (FF12N)
C2	100 nF polylayer	1(WW41U)
C3	1 μF polylayer	1(WW53H)
C4	100 nF disc ceramic	1 (BX03D)
C5	10 nF 1% polystyrene	1 (BX86T)
C6	1 nF ceramic	1 (WX68Y)

Semiconductors

TR1	BC337	1 (QB68Y)
IC1	LM331	1 (UL47B)

Miscellaneous

	8-pin DIL socket	1 (BL17T)
	pin 2145	1 (FL24B)
	PCB	1(GH20W)
	instruction leaflet	1 (XT84F)
	constructors' guide	1 (XH79L)

TDA3047 infra-red receiver

The TDA3047 is a complete infra-red receiver suitable for reception and demodulation of 100% amplitude modulated signals. Typical applications include reception of low speed data and infra-red remote control. The device includes a high frequency amplifier limiter, synchronous demodulator, AGC detector, pulse shaper and output buffer. Figure 2.33 shows the integrated circuit's block diagram and Table 2.10 shows some typical electrical characteristics for the device. The device can be used in both narrow and wide-band applications; the circuit diagram of a typical wide-band infra-red receiver is shown in Figure 2.34.

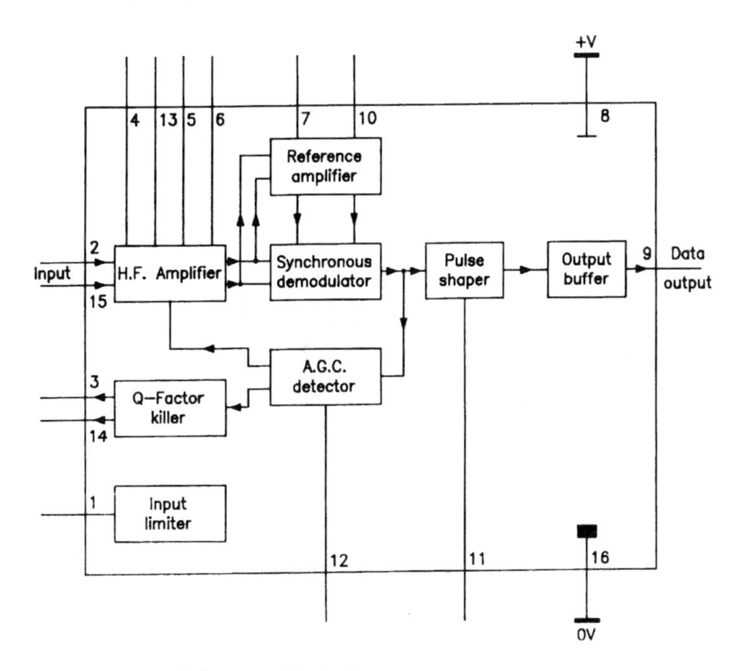

Figure 2.33 TDA3047 block diagram

Parameters	Conditions	Min	Typ	Max
Supply voltage		4.7 V		5.4 V
Supply current			2.1 mA	
Input signal	Peak to peak, 100% AM, frequency 36 kHz	0.02 mV		200 mV
Output signal	Output high	4.5 V	4.9 V	
	Output low		0.1 V	0.5 V
Output current		75 μA	120 μA	
Pulse shaper output current				10 mA
Operating temperature		–25°C		150°C

Note: above specification based on 5 V supply and ambient temperature of 25°C.

Table 2.10 Typical electrical characteristics of the TDA3047

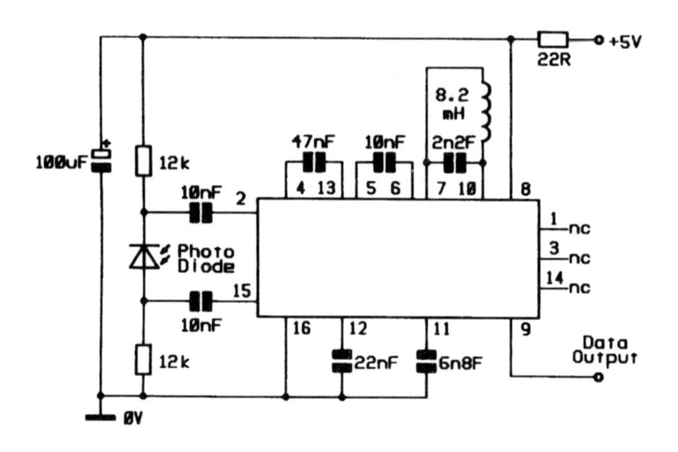

Figure 2.34 Typical wide-band infra-red receiver

Integrated circuit description

Input signals are initially fed to a high frequency (HF) amplifier. The HF amplifier consists of three d.c. amplifier stages connected in cascade to give an overall gain in the region of 83 dB. Gain control starts in the second stage of the amplifier and is then transferred to the first stage as the second stage limits; this helps to maintain an optimum signal-to-noise ratio. Two negative feedback loops are incorporated in the design to prevent excessive offset voltages in the d.c. coupled amplifier. After initial amplification the signal is applied to both the synchronous demodulator and the reference amplifier.

Signals from the HF amplifier and the reference amplifier are multiplied by the synchronous demodulator. The reference amplifier exhibits approximately 0 dB voltage gain and effectively acts as a limiting stage. The output signal from the demodulator is fed to the AGC detector and pulse shaper.

The AGC detector consists of two NPN transistors arranged as a differential pair. Peak signals from the demodulator are detected by the AGC circuit and an internal integrator capacitor removes any noise pulses. The output from the AGC detector is amplified and fed to the *Q-factor killer* and the first and second stages of the HF amplifier.

A separate differential pair connected in parallel with the AGC circuit comprises the pulse shaper. The output of the pulse shaper is determined by the voltage across the capacitor connected to pin 11 of the IC and this voltage is applied directly to the output buffer. The buffer incorporates a hysteresis circuit to protect against excessive voltage spikes at the output. Output voltage is typically 4.9 V peak to peak and is active low.

When the device is used in narrow band applications it is necessary to reduce the selectivity of the input; this is especially the case when large signals are present. The integrated circuit incorporates a *Q-factor killer* which can be directly coupled to the input for use in narrow band circuits.

IC power supply requirements

The TDA3047 operates from a 5 V d.c. power supply with a typical current drain of around 2 mA. It is important that the supply is adequately filtered to prevent the introduction of noise into the system. High frequency decoupling should be as close to the device as possible to prevent external noise pickup.

Circuit

A circuit diagram for the TDA3047 is shown in Figure 2.35 while Figure 2.36 shows a possible layout. A link option is provided so that the circuit may be used with or without on-board voltage regulation. When using the module without on-board regulation, fit link LK1 (if the circuit is being used in this configuration do not fit voltage regulator RG1); however, if on-board voltage regulation is required, RG1 should be fitted and link LK1 should be omitted. When the circuit is used without on-board regulation, the supply voltage to the module must be between 4.7 V and 5.4 V. If regulation is used, the module will operate over a wide range of supply voltages between 7 V and 30 V.

Power supply connections are made to P1 (+V) and P2 (0 V) and the demodulated data output is taken between P5 (data) and P6 (0 V). Phototransistor TR1 is connected between P3 (collector) and P4 (emitter).

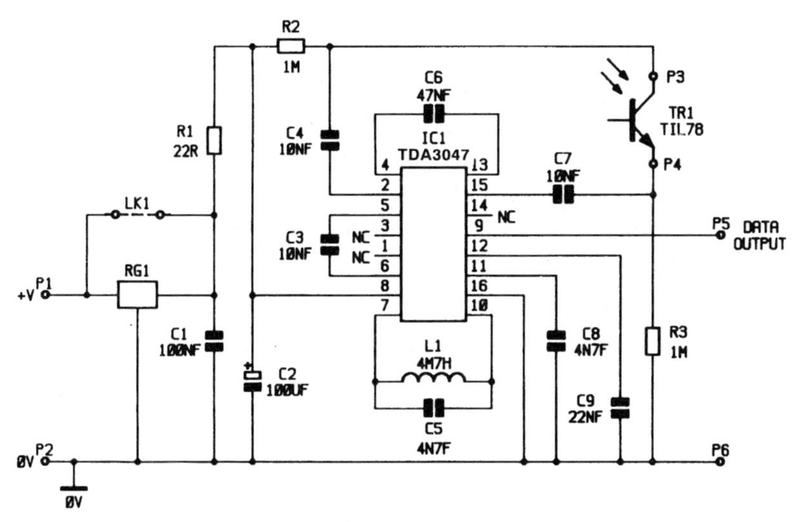

Figure 2.35 Circuit diagram

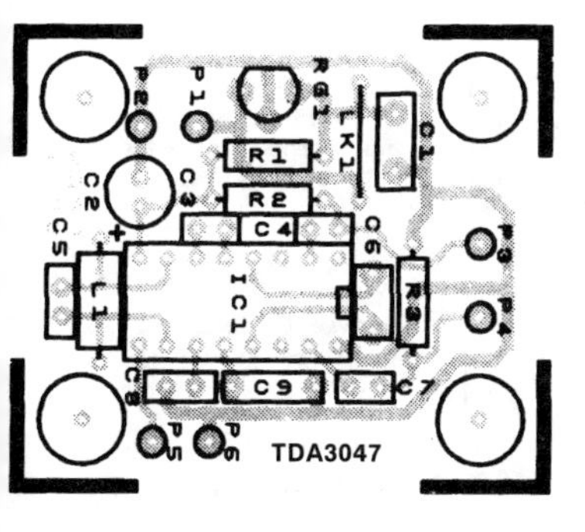

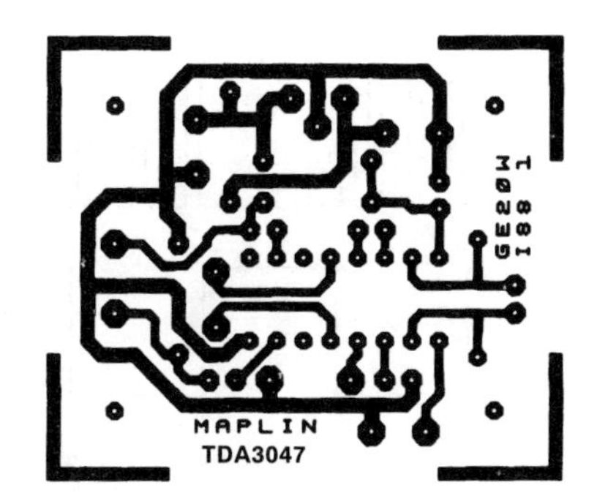

Figure 2.36 PCB layout

The circuit is suitable for use with 100% amplitude modulated signals (a pulse or squarewave modulating a carrier). A typical example of a suitable transmission format is shown in Figure 2.37. If a simple carrier is received (without modulation) the data output will change from high to low and remain in this state as long as the transmission is being received.

The operating frequency of the circuit is approximately 36 kHz using the component values shown in the Parts List. Operating frequency is basically determined by the resonant frequency of the tuned circuit formed by inductor L1 and capacitor C5; the corresponding L/C ratio determines the bandwidth of the receiver. Using the values shown the receive bandwidth is relatively wide and the circuit will accommodate carrier frequencies between approximately 30 kHz and 40 kHz with a

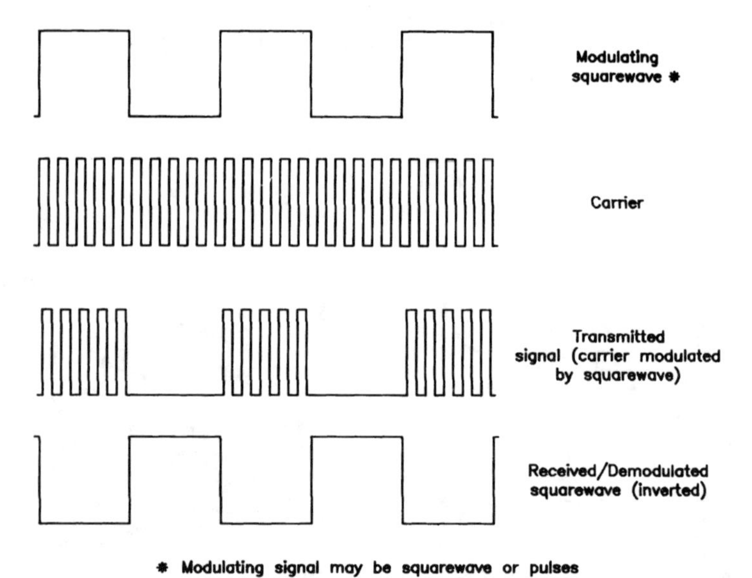

Figure 2.37 Typical example of a suitable transmission format

peak in performance around the centre of this range. Receive frequency and bandwidth can be tailored to suit different applications by changing the values of inductor L1 and capacitor C1.

For optimum performance, the TDA3047 is best driven into a high impedance, as the output current is limited to around 120 μA; the device can of course drive a lower impedance with a reduction in output voltage. The TDA3047 provides an active low output. Table 2.11 shows the specification of the prototype circuit built using the earlier printed circuit board while Figure 2.38 shows the printed circuit board wiring diagram.

Power supply voltage	With on-board regulation	7 V–30 V
	Without regulation	4.7 V–5.4 V (5 V nominal)
Current consumption	(12 V power supply, RG1 fitted)	5.2 mA
	(5 V power pupply, LK1 fitted)	1.98 mA
Receive frequency (carrier)		Approximately 36 kHz

Table 2.11 Specification of prototype (built using the PCB)

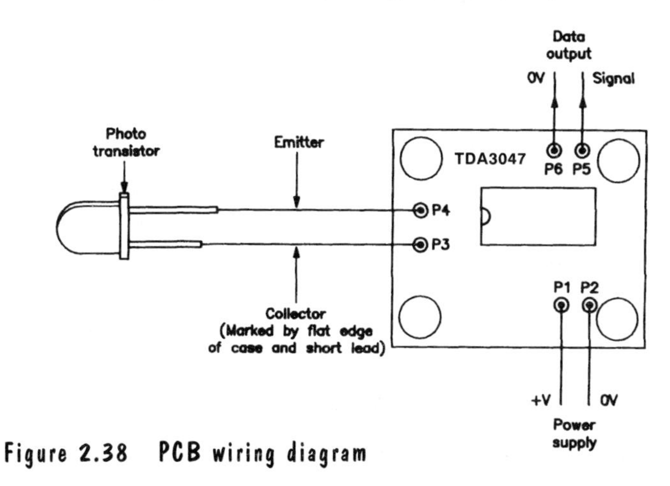

Figure 2.38 PCB wiring diagram

TDA 3047 Infra-red receiver parts list

Resistors — All 1% 0.6 W metal film

R1	22 Ω	1	(M22R)
R2,3	1 MΩ	2	(M1M)

Capacitors

C1	100 nF minidisc	1	(YR75S)
C2	100 μF 10 V PC electrolytic	1	(FF10L)
C3,4,7	10 nF ceramic	3	(WX77J)
C5,8	4n7F ceramic	2	(WX76H)
C6	47 nF minidisc	1	(YR74R)
C9	22 nF polylayer	1	(WW33L)

Semiconductors

TR1	TIL78	1	(YY66W)
RG1	78L05AWC	1	(QL26D)
IC1	TDA3047	1	(UL25C)

Miscellaneous

L1	4.7 mH choke	1	(UK80B)
	16-pin DIL socket	1	(BL18V)
	pins 2145	1	(FL24B)

3 Audio projects

LM1037 analogue switch

The LM1037 is a dual, four channel, analogue switch incorporating an internal muting facility. The device is suitable for a wide range of switching applications including multiplexing and stereo source selection. Each channel is selected by one of four control inputs. Figure 3.1 shows the integrated circuit pin-out diagram and Table 3.1 and Figure 3.2 show typical electrical characteristics for the LM1037.

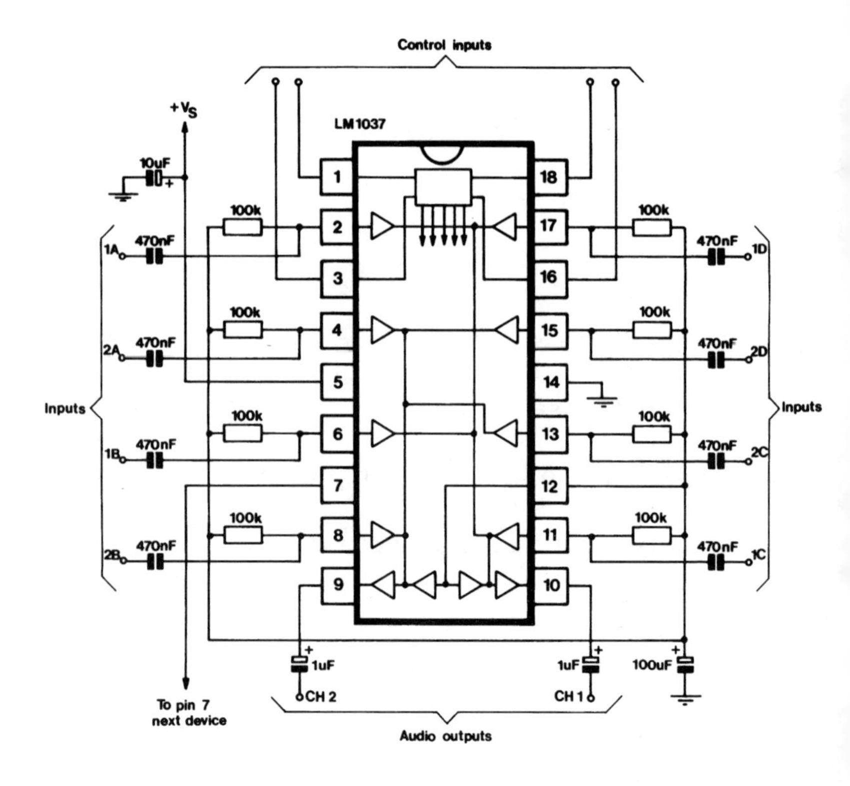

Figure 3.1 IC pin-out diagram

Parameter	Operating conditions	Min	Typ	Max
Supply voltage		5 V		28 V
Supply current	Supply voltage = 12 V		6.4 mA	8.5 mA
Maximum input voltage	Supply voltage = 12 V	2.8 V	2.9 V	3.0 V
Operating temperature		−20°C		70°C
Total harmonic distortion (THD)	Signal input = 1 V r.m.s. at 1 kHz		0.04%	0.1%
Voltage gain		−0.5 dB	0 dB	0.5 dB
Channel separation	Signal input = 1 V r.m.s. at 1 kHz		−95 dB	
Control voltage	High level	2 V		50 V
	Low level			0.8 V

Note: above specifications assume a 12 V power supply and an ambient temperature of 25°C.

Table 3.1 Typical electrical characteristics of LM1037

Channel selection

Channel selection is achieved by applying a d.c. voltage of between 2 V and 50 V (maximum) to one of four separate control pins. Each control pin selects a different input channel. It is possible to switch an increased number of channels, using two integrated circuits by connecting the mute inhibit pins (pin 7) of the devices together and connecting the two pairs of output pins (pin 9 and pin 10) in parallel. It should be noted that only one output capacitor is required for each channel.

Input signals should be limited to approximately 2.5 V r.m.s. (for a 12 V power supply) to avoid excessive distortion. The maximum instantaneous voltage between

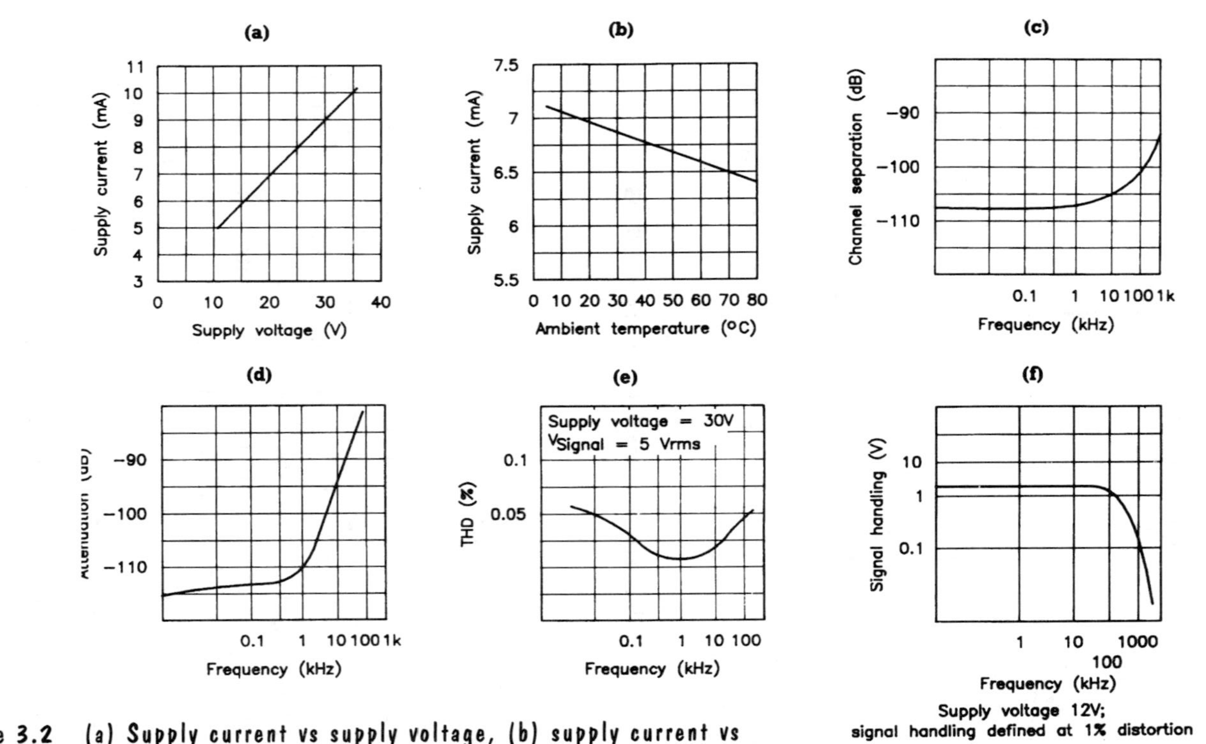

Figure 3.2 (a) Supply current vs supply voltage, (b) supply current vs temperature, (c) channel separation vs frequency, (d) attenuation of unselected inputs vs frequency, (e) THD vs frequency, (f) signal handling vs frequency.

the two inputs of any one channel is 9.6 V; voltages in excess of this value may cause degraded channel separation or increased distortion. It should be noted that the maximum signal handling is dependent on the power supply voltage (at higher supply voltages the device is capable of handling larger signals).

Power supply requirements

The LM1037 will operate over a wide range of power supply voltages between approximately 8 V and 28 V. Integrated circuit current drain is typically around 6 mA at 12 V. It is necessary to decouple the supply rail close to the integrated circuit to avoid the introduction of excessive noise and to prevent possible instability.

Printed circuit board

A high quality fibreglass printed circuit board with printed legend is available for the basic LM1037 four channel stereo source selector circuit. Figure 3.3 shows the circuit diagram and Figure 3.4 shows the printed circuit board layout. A single supply of between 8 V and 28 V is required to power the circuit. The power supply should be capable of delivering at least 15 mA and should be adequately decoupled to prevent the introduction of mains derived noise onto the supply rail. Power supply connections are made to P1(+V) and P2(0 V). All connections are illustrated in Figure 3.5. Channel selection is

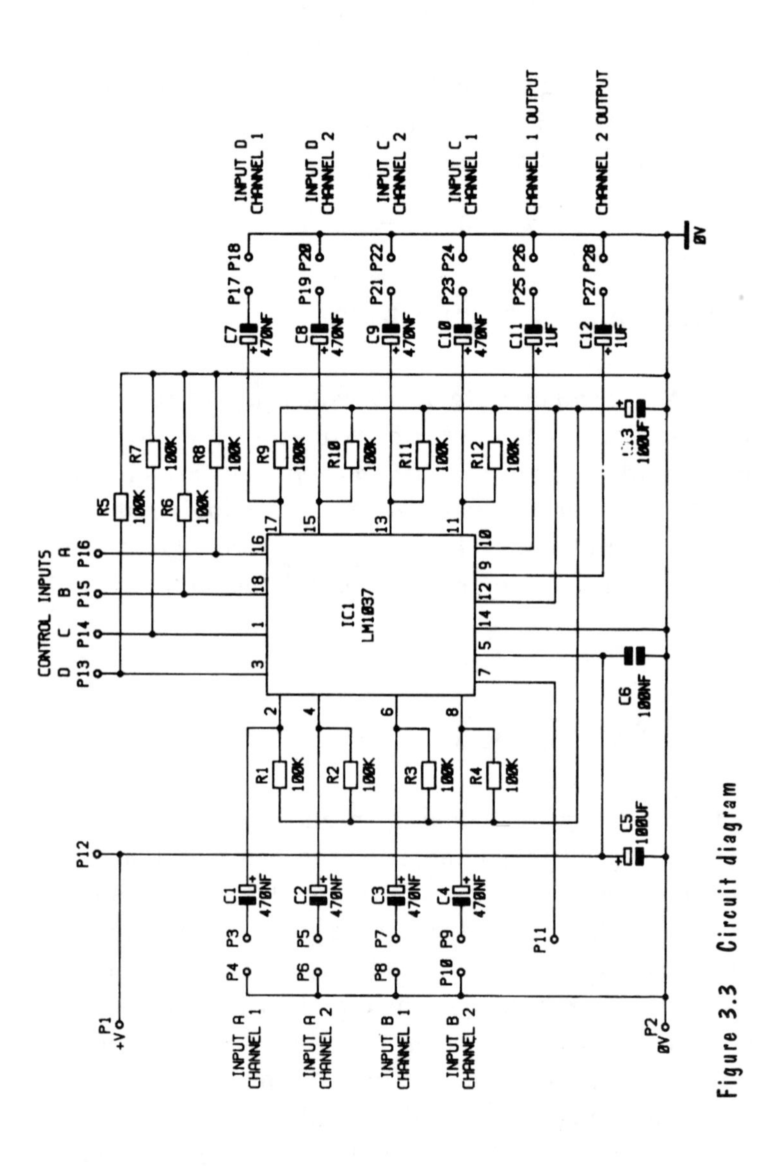

Figure 3.3 Circuit diagram

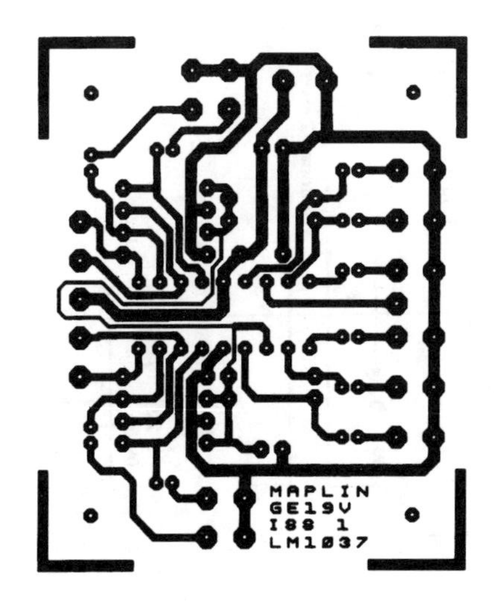

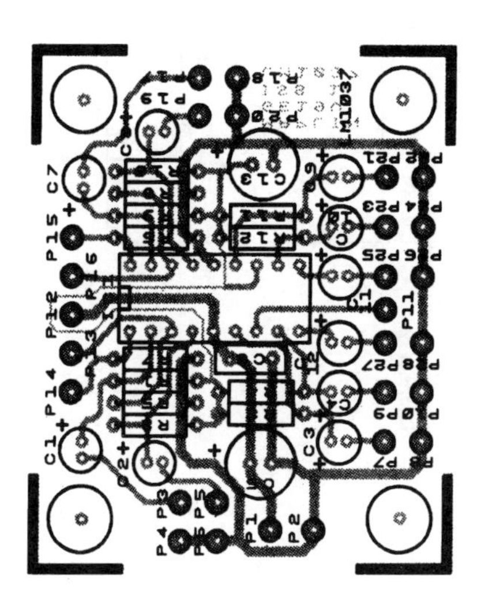

Figure 3.4 PCB layout

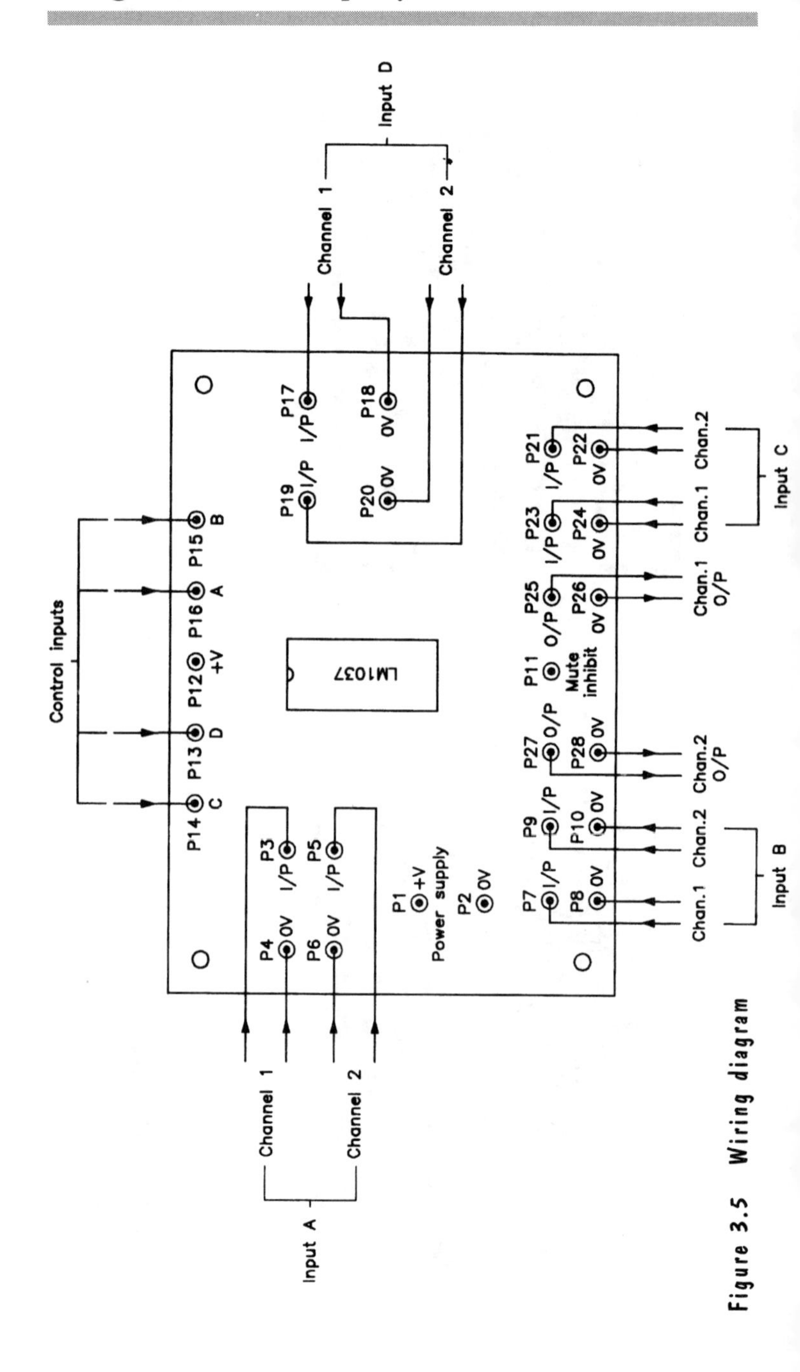

Figure 3.5 Wiring diagram

achieved by applying a d.c. voltage of between 2 V and 50 V to P13, P14, P15 or P16 as appropriate (refer to Table 3.2). Access to the mute inhibit line is provided by P11. Finally Table 3.3 shows the specification for the prototype circuit built using the printed circuit board.

		Inputs switched to output pins							
Input channel		1A	2A	1B	2B	1C	2C	1D	2D
Pin number		P3	P5	P7	P9	P23	P21	P17	P19
Control pins	P13	Low	Low	Low	Low	Low	Low	High	High
	P14	Low	Low	Low	Low	High	High	Low	Low
	P15	Low	Low	High	High	Low	Low	Low	Low
	P16	High	High	Low	Low	Low	Low	Low	Low
Output pin		P25	P27	P25	P27	P25	P27	P25	P27
Output channel		1	2	1	2	1	2	1	2

Table 3.2 Truth table for LM1037 module

Supply voltage range	8 V–18 V
Supply current (12 V)	5.5 mA
Maximum input voltage (12 V supply)	3 V r.m.s.
Voltage gain	0 dB
THD (input voltage = 770 mV r.m.s. at 1 kHz)	0.02%

Table 3.3 Specification of prototype (built using the PCB)

LM1037 parts list

Resistors — All 1% 0.6 W metal film

R1–12	100 k	12	(M100K)

Capacitors

C1–4, 7–10	470 nF 100 V PC electrolytic	8	(FF00A)
C5,13	100 μF 35 V PC electrolytic	2	(JL19V)
C6	100 nF disc	1	(BX03D)
C11,12	1 μF 100 V PC electrolytic	2	(FF01B)

Semiconductors

IC1	LM1037N	1	(QY33L)

Miscellaneous

	constructors' guide	1	(XH79L)
	printed circuit board	1	(GE19V)
	18-pin DIL socket	1	(HQ76H)
	pins 2145	1	(FL24B)

The above parts are available as a kit, order as LP06G

SSM2015 low noise pre-amplifer

The SSM2015 is a low noise pre-amplifier integrated circuit featuring low distortion and a wide bandwidth. Voltage gains between approximately 10 and 2000 can be set using different resistor values and the device is ideal for microphone pre-amplification. True differential inputs make the integrated circuit particularly useful for interfacing balanced transducers to equipment with single ended inputs. Figure 3.6 shows the integrated circuit pin-out diagram and Table 3.4 shows typical electrical characteristics for the SSM2015.

Circuit description

The SSM2015 is a true differential amplifier with the feedback return path directly to the emitters of the input stage transistors. Using this system, it is possible to obtain a very good noise figure and high common mode rejection. The input stage current is maintained by an internal feedback loop which is in turn controlled by an external bias resistor allowing programmability and also enabling the noise figure to be optimised for a wide range of source impedances

Figure 3.7 shows a typical application circuit for the SSM2015. Overall frequency compensation for the circuit is provided by capacitors C1 and C2 while capacitor C3 compensates the input stage current regulator. The values of capacitors C1 and C2 are dependent on the value of bias resistor, R_{bias} and typical values range from 15 pF to 30 pF for capacitor C1 and 5 pF to 15 pF for capacitor C2.

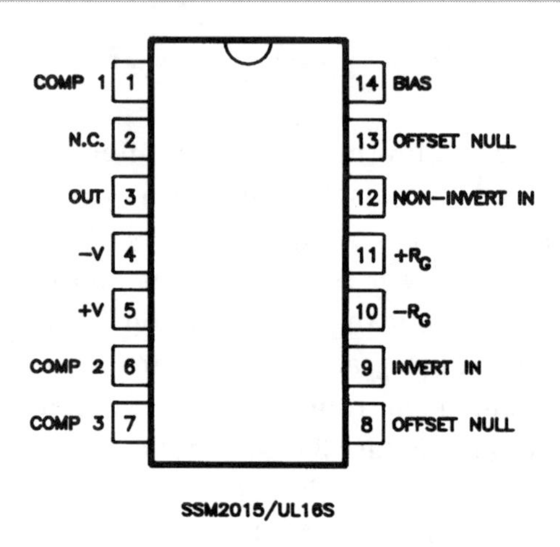

Figure 3.6 IC pin-out

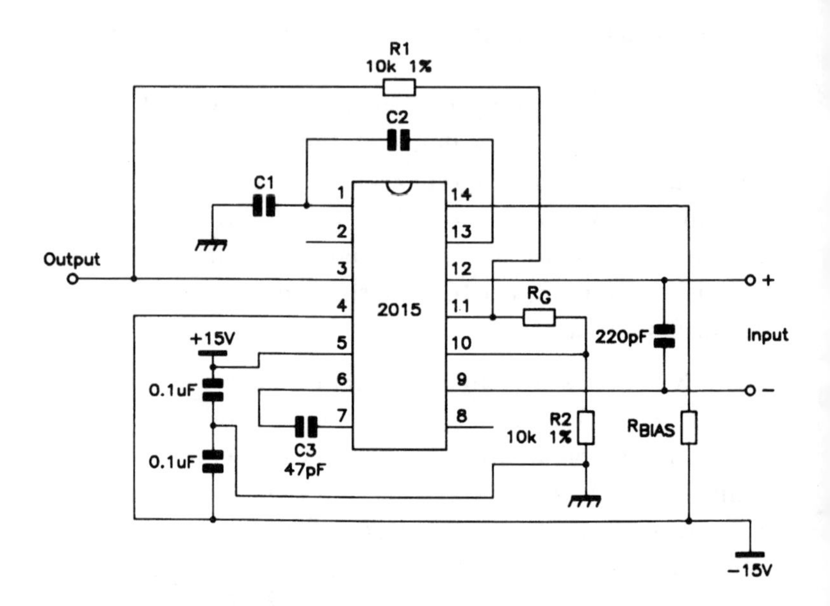

Figure 3.7 Typical application circuit

Parameter	Conditions	Min	Typ	Max
Power supply voltage (d.c.)		±12 V	±15 V	±17 V
Supply current	±15 V		12 mA	16 mA
Total harmonic distortion (THD)	Output voltage = 7 V r.m.s. Gain = 1000, Frequency = 1 kHz		0.007%	0.01%
	Output voltage = 7 V r.m.s. Gain = 100, Frequency = 1 kHz		0.007%	0.01%
	Output voltage = 7 V r.m.s. Gain = 10, Frequency = 1 kHz		0.01%	0.015%
Output current	Source	15 mA	25 mA	
	Sink	8 mA	14 mA	
Bandwidth (–3 dB)	Gain = 10		1 MHz	
	Gain = 100		700 kHz	
	Gain = 1000		150 kHz	
Output voltage swing	2 kΩ load	±10.5 V	±12.5 V	
Slew rate			6 V/μs	
Common mode input impedance			50 MΩ	
Differential mode Input impedance	Gain = 10		20 MΩ	
	Gain = 100		5 MΩ	
	Gain = 1000		0.5 MΩ	
Common mode voltage range		±4 V	±5 V	

Above specification for ±15 V power supply, temperature = 25°C.

Table 3.4 SSM2015 typical electrical characteristics

The bandwidth of the SSM2015 is dependent on gain and the value of the bias resistor (R_{bias}). Under worst case conditions (gain = 1000, R_{bias} = 150 k) the bandwidth is in excess of 70 kHz; however, at lower gains and higher bias currents, –3 dB bandwidths up to 1 MHz are possible.

The SSM2015 inputs are completely floating and, to prevent the input voltage from exceeding the common mode range, it is necessary to provide a d.c. connection to a suitable point in the circuit (usually 0 V). A common method of achieving this is to connect one side of the input transducer to 0 V. Another method is to let the transducer float and connect each input to the 0 V line using resistors; the value of these should be kept as low as possible for noise immunity purposes and in practice may be any value up to a maximum of around 10 k. For optimum noise performance balanced transducers should be used and these may be connected directly to the integrated circuit inputs without additional bias resistors.

Kit available

A kit of parts using a high quality, fibreglass printed circuit board is available for a basic application circuit using the SSM2015. Figure 3.8 shows the circuit diagram of the module and Figure 3.9 shows the legend of the printed circuit board. The circuit will handle either balanced or single ended inputs and additional components are used to give a balanced output as well as the standard single ended output.

A split rail power supply of between ±12 V and ±17 V is required to power the module. The supply should be capable of delivering at least 50 mA and should be properly smoothed and regulated. Connection information for the module is shown in Figure 3.10.

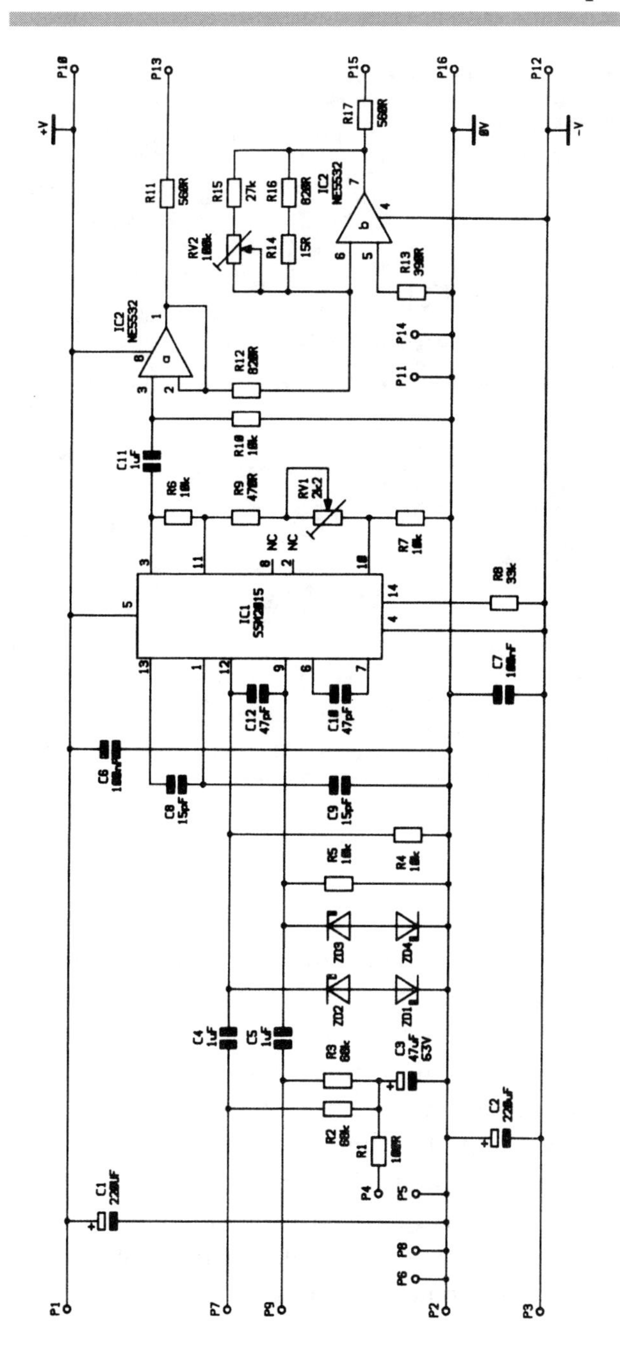

Figure 3.8 Circuit diagram

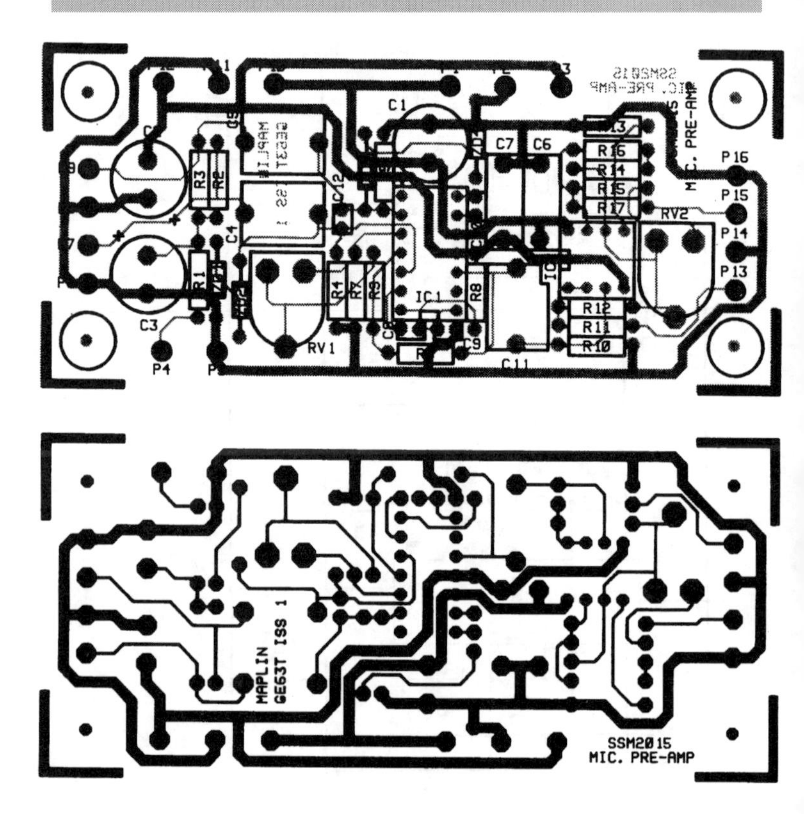

Figure 3.9 Legend

Presets RV1 and RV2 provide adjustable gain. The overall gain of the pre-amplifier is adjusted by preset RV1 and may be set between approximately 10 and 50. Maximum gain is achieved when preset RV1 is set to the fully anti-clockwise position. Preset RV2 allows fine adjustment of the gain of IC2b either side of 0 so that this can be matched to the gain of IC2a for correct balancing. If no test equipment is available, preset RV2 should be set to the centre of its travel; although this does not necessarily provide optimum matching of the two output

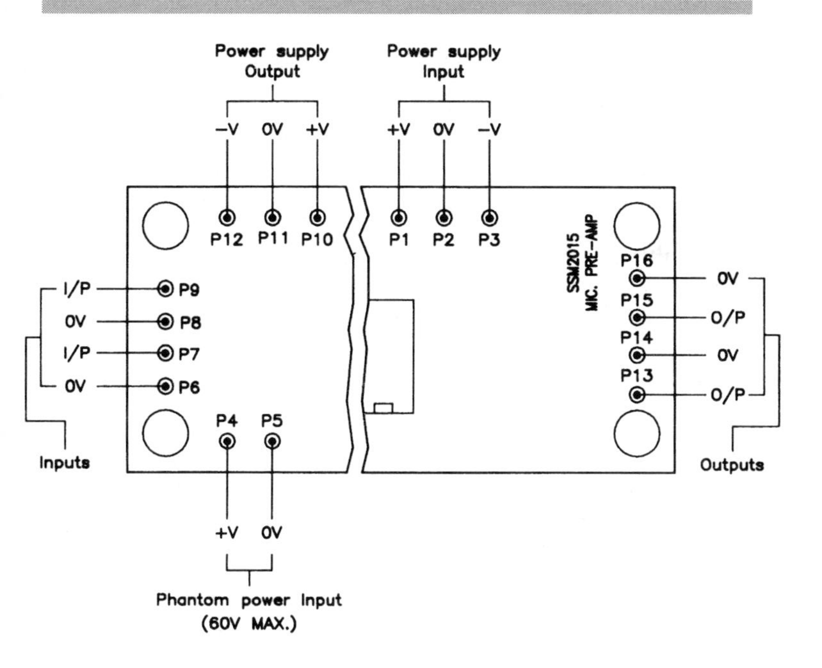

Figure 3.10 Wiring diagram

levels, a reasonable degree of balance should be obtained. Balanced inputs are connected to P7(+i/p), P8(0 V) and P9(–i/ p). If a single ended output is required this may either be taken between P13(o/p) and P14(0 V) or between P15(o/p) and P16(0 V). For balanced operation, output connections are made to P13(o/p), P14(0 V) and P15(o/p). Note that the output at P13 is 180 degrees out of phase with the output at P15.

Additional facility is provided to allow a low current power supply up to a maximum of 60 V to be coupled onto the input lead for phantom powering microphones and similar equipment. Zener diodes, ZD1–ZD4 provide input protection for the SSM2015 from any voltage transients that may occur when connecting microphones/ microphone power supplies to the module. If the phantom power facility is not required, then the components used in this part of the circuit (resistors R1, R2, and R3, and capacitor C3) need not be fitted.

Applications

The pre-amplifier module may be used in a wide range of applications but is particularly useful when interfacing balanced transducers (microphones) to single ended equipment. Balanced connections are normally made using XLR connectors and the standard wiring format for this type of connector is shown in Figure 3.11. Stereo

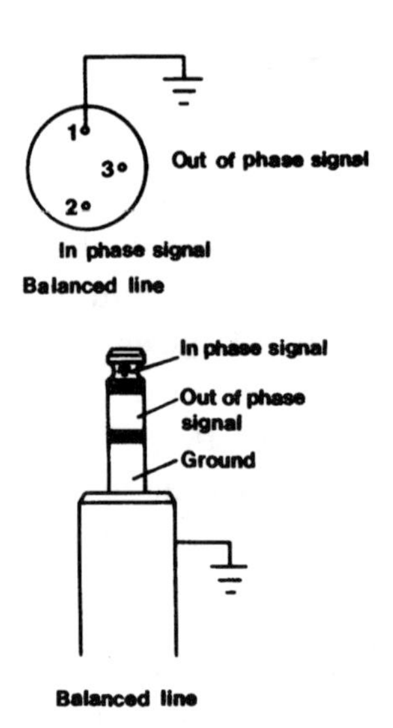

Figure 3.11 Standard XLR and jack connections for balanced line

jack connectors may also be used and typical connections for these are also shown. Finally, three additional pins P10(+V), P11(0 V) and P12(–V) are included for low current (500 mA maximum) power supply connections to auxiliary equipment; the pins are connected electrically via tracks on the printed circuit board to P1, P2 and P3. Table 3.5 shows the specification of the prototype pre-amplifier.

Parameter		Value
Supply voltage		±12 V to ±17 V d.c.
Supply current (quiescent)		20 mA at ±15 V
Total harmonic distortion (THD) measured at 1 kHz		Less than 0.02%
Gain:	RV1 set to minimum gain	21.0 dB
	RV1 set to maximum gain	33.5 dB
Maximum input voltage (for 9 V r.m.s. output, gain set to maximum, ±15 V power supply)		190 mV r.m.s.
Maximum input voltage (for 9 V r.m.s. output, gain set to minimum, ±15 V power supply)		800 mV r.m.s.
PCB dimensions		43 mm x 99 mm

Table 3.5 Specification of prototype module

SSM2015 microphone pre-amplifier parts list

Resistors — All 0.6 W 1% metal film

R1	100 R	1 (M100R)
R2,3	68 k	2 (M68K)
R4,5,6, 7,10	10 k	5 (M10K)
R8	33 k	1 (M33K)
R9	470 R	1 (M470R)
R11,17	560 R	2 (M560R)
R12,16	820 R	2 (M820R)
R13	390 R	1 (M390R)
R14	15 R	1 (M15R)
R15	27 k	1 (M27K)
RV1	2k2 hor enc preset	1 (UH01B)
RV2	100 k hor encl preset	1 (UH06G)

Capacitors

C1,2	220 μF 35 V PC electrolytic	2 (JL22Y)
C3	47 μF 63 V PC electrolytic	1 (FF09K)
C4,5,11	1 μF polylayer	3(WW53H)
C6,7	100 nF polyester	2 (BX76H)
C8,9	15 pF ceramic	2 (WX46A)
C10,12	47 pF ceramic	2 (WX52G)

Semiconductors

IC1	SSM2015	1 (UL16S)
IC2	NE5532	1 (UH35Q)
ZD1–4	BZX61C6V2	4 (QF48C)

Miscellaneous

P1–16	pin 2145	1	(FL24B)
	8-pin DIL socket	1	(BL17T)
	14-pin DIL socket	1	(BL18U)
	printed circuit board	1	(GE63T)
	constructors' guide	1	(XH79L)

The above parts are available as a kit, order as LP42U

SL6270 AGC microphone pre-amplifier

The SL6270 is a small 8-pin integrated circuit combining the functions of an audio amplifier and Voice Operated Gain Adjusting Device (VOGAD). It is designed to accept small signals from a microphone and to provide an essentially constant output signal from an input covering a range of 50 dB. The dynamic range, attack and decay times are controlled by external components. The device will operate over a wide range of power supply voltages between 4.5 V to 10 V and consumes only 9 mA from a 9 V battery. Figure 3.12 shows the IC pin-out, and Table 3.6 gives the electrical characteristics of the device.

Test conditions — supply voltage V_{CC}: 6 V
input signal frequency: 1 kHz
ambient temperature: –30°C to +85°C

Characteristic	*Value*			*Units*	*Conditions*
	Min	*Typ*	*Max*		
Supply current		5	10	mA	
Input impedance		150		Ω	Pin 4 or 5
Differential input impedance		300		Ω	
Voltage gain	40	52		dB	72 μV r.m.s. input pin 4
Output level	55	90	140	mV r.m.s.	4 mV r.m.s. input pin 4
THD		2	5	%	90 mV r.m.s. input pin 4
Equivalent noise input voltage		1		μV	300 Ω source, 400 Hz to 25 kHz bandwidth

Table 3.6 Typical electrical characteristics

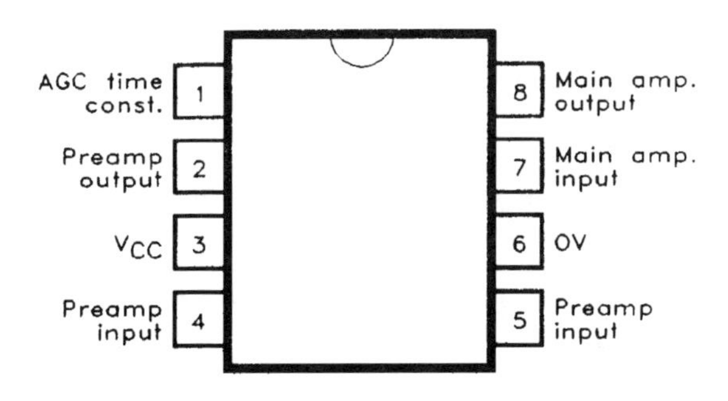

Figure 3.12 IC pin connections

IC description

Figure 3.13 shows the block diagram of the SL6270. Its positive power supply input is on pin 3, which should be decoupled to 0 V ground by a small 100 nF ceramic capacitor. Pin 6 is used as a common 0 V ground return for all stages within the device.

The AGC controlled pre-amplifier stage has a true differential input on pins 4 and 5, allowing it to be driven single endedly without the problems caused by other forms of push-pull circuits. The applied audio signal must be a.c. coupled to the input via a capacitor, and in the single ended mode it can be applied to either input. Each input has a very low impedance, only 150 Ω if single-ended, doubling up to 300 Ω if differential mode is used. Signals of less than a few hundred microvolts are amplified normally, but as the input level increases the AGC begins to

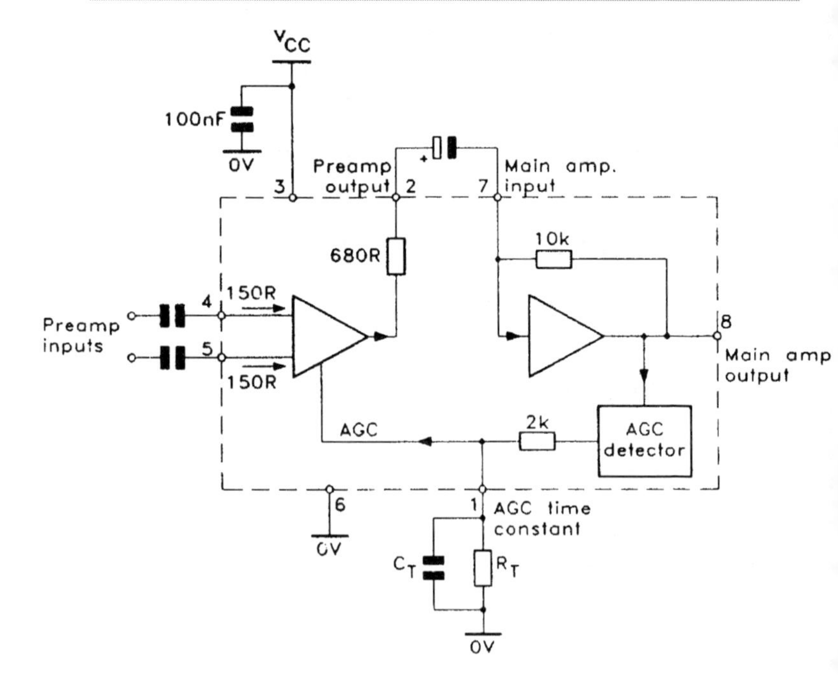

Figure 3.13 Block diagram

take effect and the output signal on pin 8 is held constant at approximately 90 mV r.m.s. over the full input range of 50 dB, see Figure 3.14. The test conditions used to obtain this graph were as follows: power supply voltage +6 V, input frequency 1 kHz, ambient temperature +25°C, single-ended input. In order to ensure that internal offsets within the amplifier are of such polarity as to inhibit oscillation at the onset of AGC, a 22 k resistor should be connected from pin 5 to the 0 V supply line.

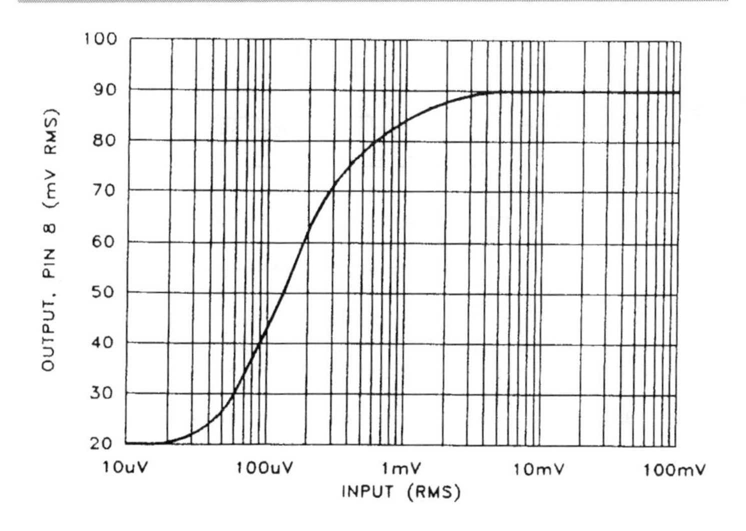

Figure 3.14 Input/output characteristics

The output from the AGC pre-amp stage (pin 2) is coupled via a capacitor to the input of the main amplifier (pin 7). The low frequency (LF) response of the system is determined by the value of this capacitor and the internal 680 Ω resistor on pin 2. For normal speech radio communications, this coupling capacitor is chosen to give a –3 dB output amplitude point at 300 Hz, which corresponds to 2.2 μF. The LF response can be extended down to 100 Hz or less by simply increasing the value of this capacitor, although values greater than 100 μF are not recommended.

The combined amplification system has an upper frequency response which extends beyond 1 MHz, see Figure 3.15. However, this bandwidth can be restricted by placing an external capacitor between pins 7 and 8.

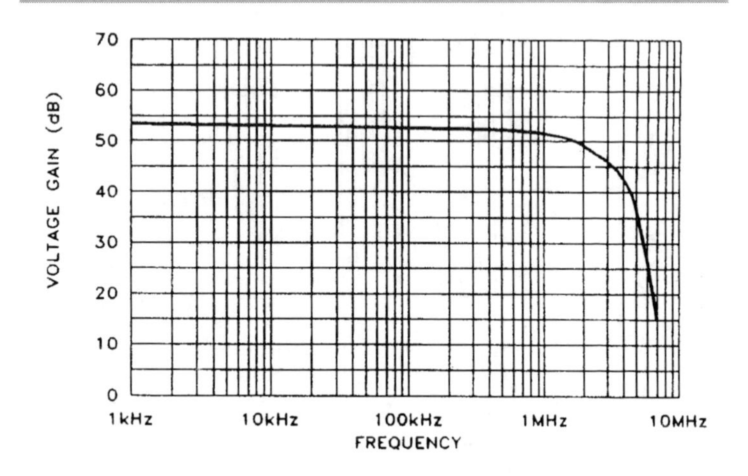

Figure 3.15 Frequency response

These are the input and output connections of the main amplifier, which has an internal 10 kΩ resistor to set the flat response open loop gain. As the value of the external capacitor increases, the upper frequency response decreases and for radio communications this is normally chosen to give a –3 dB point at 3 kHz, which corresponds to 4.7 nF. In addition, the dynamic range and sensitivity of the system can be reduced by placing a resistor between these two pins, and as its value decreases, the voltage gain of the main amplifier will continue to drop. An approximate reduction of 20 dB in gain can be achieved with a value of 1 kΩ, but values less than 680 Ω are not advised. The final audio output on pin 8 has a small d.c. offset voltage, so it is good practice to use an a.c. coupling capacitor before connecting it to any other circuits. To ensure a good LF response into loads as low as 1 kΩ, this capacitor should be not less than 22 μF

The main amplifier also drives the AGC detector, and the d.c. voltage generated by this circuit is directly related to the signal level applied to the pre-amp input pins 4 and 5. This control voltage is used to progressively reduce the gain of the pre-amp as the input level increases. Because of its careful design, the device has an extremely high input dynamic range and using the same test conditions as before; Figure 3.16 shows the distortion characteristics at various input levels.

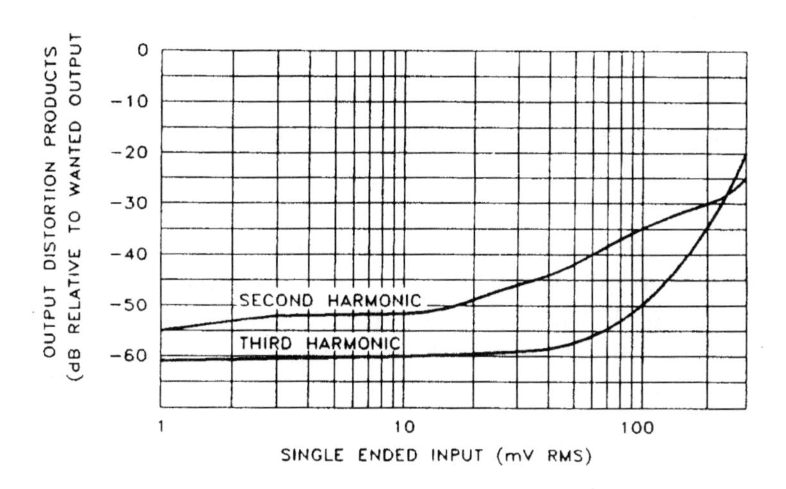

Figure 3.16 Distortion products versus input signal

The AGC attack and decay characteristics are set by the external RC timing components on pin 1. Normally the SL6270 is required to respond quickly by holding the output level almost constant as the input level is increased. This attack time is set by the value of the capacitor C4, and is defined as the time taken for the output to return to within 10% of the original level following a 20 dB increase in input level. A fast attack response can be obtained by using an electrolytic capacitor in the range 22 μF to 47 μF. The decay time is set by the value of the resistor R6, which discharges the current held by the capacitor, the recommended decay rate being approximately 20 dB/second, and this slower response time is obtained by using a value of 1 MΩ.

Maplin kit

A kit of parts, including a small fibreglass PCB with printed legend, is available as stock code LP98G. The basic kit as supplied is for use with radio communication equipment, and Table 3.7 shows the specification of the prototype. Because the SL6270 may be used in many varied applications, some of the component values supplied in the basic kit will not be suitable, so alternative values must be calculated to determine the new working parameters.

Power supply voltage:	+4.5 V to +10 V (P6) and 0 V (P5)
Power supply current:	11 mA (+9 V supply)
Microphone input:	electret or dynamic inserts
Input impedance:	electret 4k7Ω (P1) and 0 V (P2)
	dynamic 150 Ω (P3) and 0 V (P2)
	dynamic 300 Ω (P3 and P4)
Voltage gain:	52 dB
Maximum output:	90 mV r.m.s. (P9) and 0 V (P10)
Minimum load impedance:	1 kΩ
Frequency bandwidth (–3 dB):	300 Hz to 3 kHz

Table 3.7 Specification of prototype

Figure 3.17 shows the circuit diagram of the module, and as can be seen, some additional components have been included to provide the following extended features. A low noise transistor, TR1, is used to provide a suitable input for an electret microphone insert which is included in the kit. A preset resistor, RV1, provides the means of adjusting the audio output level. Finally, terminal pins are provided for the offboard connections:

- P1 — electret microphone insert signal input,
- P2 — electret microphone insert 0 V ground,
- P3 — dynamic microphone insert signal input 1,
- P4 — dynamic microphone insert signal input 2,
- P5 — 0 V ground power supply (–V battery),
- P6 — +V power supply input (+4.5 V to +10 V),
- P7,P8 — dynamic range and sensitivity reduction (4k7 or 1 k resistor),
- P9 — audio output (90 mV r.m.s. into 1 k),
- P10 — audio output 0 V ground.

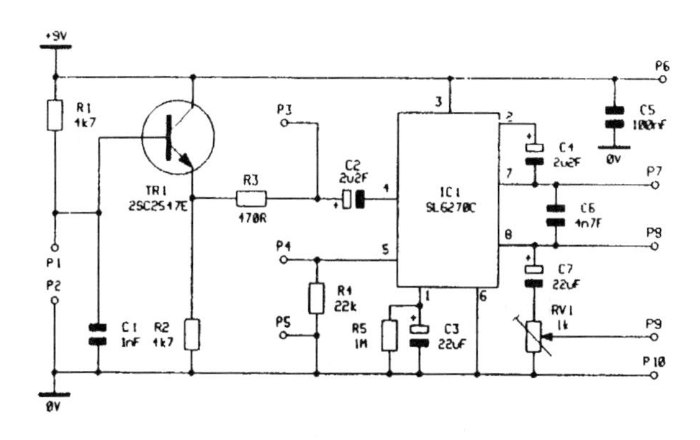

Figure 3.17 Circuit diagram

All the components and their relative positions are shown in Figure 3.18.

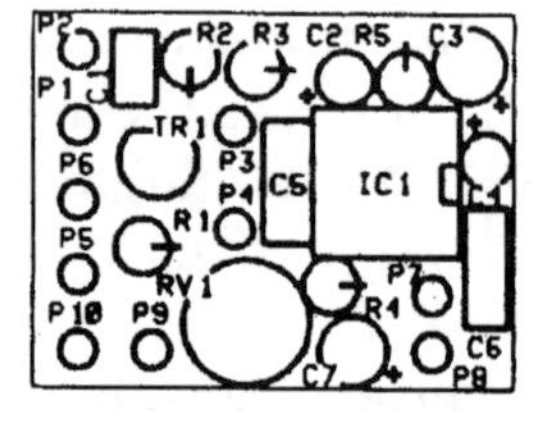

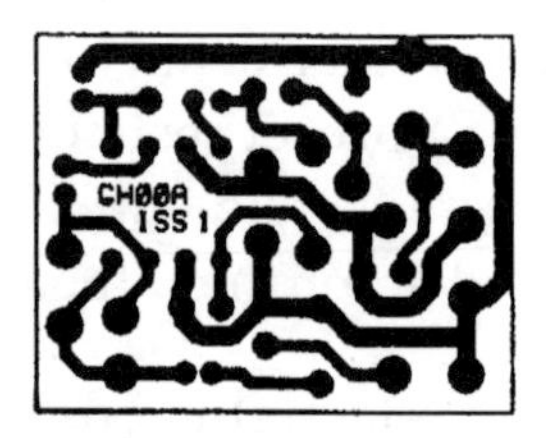

Figure 3.18 PCB legend and track

Using the module

The electret microphone insert connections to the module are illustrated in Figure 3.19(a). This wiring diagram also shows the power supply, audio output and optional dynamic range/sensitivity reduction resistor. If you intend to use a low impedance dynamic insert *(not supplied in the kit)* then the option outlined in Figure 3.19(b) (single-ended), or 3.26(c) (differential) should be followed, taking note of the necessary component changes to the PCB:

- remove R1, R2, R3, C1 and TR1,
- reverse the polarity of C2.

To minimise any stray electrical noise pick-up, the connecting leads between the microphone insert and the module should be kept as short as possible, e.g. less than

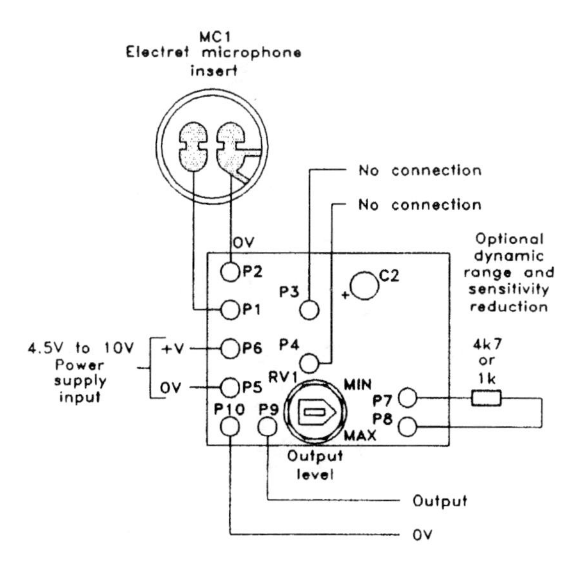

Figure 3.19(a) Wiring diagram A

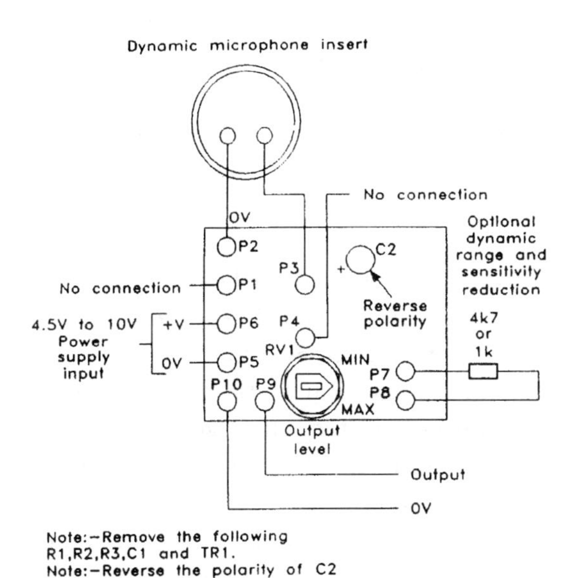

Figure 3.19(b) Wiring diagram B

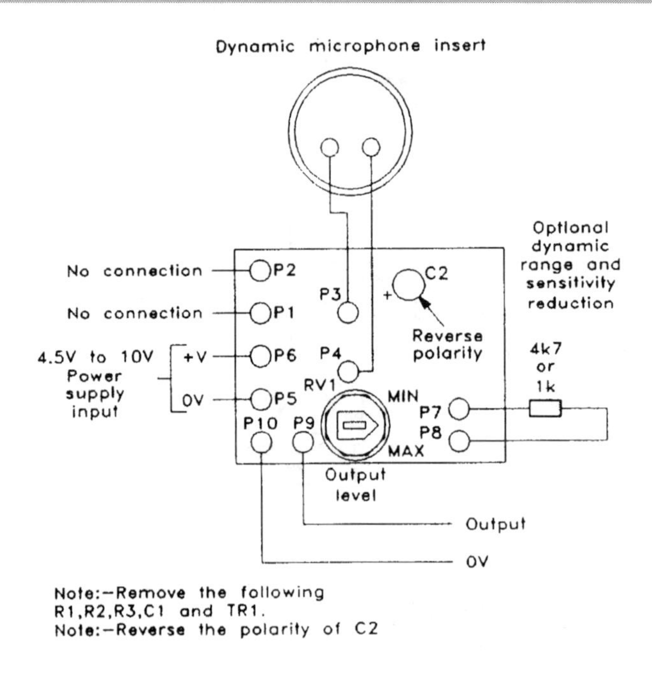

Figure 3.19(c) Wiring diagram C

50 mm. For longer runs screened cable is recommended *(as supplied in kit)* but even this should be kept as short as possible, e.g. less than 500 mm. When using electret or dynamic single-ended inserts, miniature single core screened cable is suitable. However, when using a dynamic microphone in the differential input mode, as in Figure 3.19(c), open wires or twin overall screened cable can be used. To ensure effective screening, the outer braiding of the cable should *always* be connected to the 0 V ground on P2.

The module is designed to operate over a wide supply voltage range and has a relatively low power consumption, making it suitable for battery operation. A good quality alkaline 9 V PP3-sized battery is capable of running the circuit for a considerable period of time. For this reason a PP3-sized battery clip is also included in the kit. However, no on/off switch is supplied because of the numerous switch styles, types and switching methods available, here are just a few to choose from:

Toggle, rocker, rotary, slide, push and microphone PTT.

On most radio communication microphones an integral push-to-talk (PTT) switch is used to activate the transmitter. If this switch has a spare set of contacts then the power to the module can only be applied during the transmit period, so this technique extends the operational life of the battery. If you do not intend to use a battery then it is important that the power supply is adequately decoupled to prevent audio, digital, or mains derived hum and noise from entering the circuit via the supply rail.

Screened cable should also be used on the audio output of the module and be kept as short as possible, e.g. less than 3 metres. To ensure effective screening the outer braiding of the cable must be connected to the 0 V ground at P10. The output level of the module is adjusted by RV1, with its minimum setting at the fully counter-clockwise position, see Figure 3.19.

With no additional resistor placed across pins P7 and P8 the sensitivity of the module will be at maximum. Under this condition any low level sound picked up by the microphone will be highly amplified. If there is an unduly high level of distracting background noise the overall speech intelligibility will suffer. To minimise this effect the sensitivity of the module must be reduced by adding a resistor across P7 and P8, see Figure 3.19. As its value decreases, so also will the sensitivity, until the permitted 680 Ω minimum is reached. After some experimentation the following resistor values where chosen in the prototype:

- no resistor = full sensitivity,
- 4k7 resistor = medium sensitivity,
- 1 k resistor = low sensitivity.

As the sensitivity decreases it will become necessary to speak louder, and/or closer to the microphone in order to maintain the same average output level.

AGC microphone pre-amplifier parts list

Resistors — All 0.6 W 1% metal film (unless specified)

R1,2	4k7	2	(M4K7)
R3	470 Ω	1	(M470R)
R4	22 k	1	(M22K)
R5	1 M	1	(M1M)
RV1	1 k cermet	1	(WR40T)
	4k7	1	(M4K7)
	1 k	1	(M1K)

Capacitors

C1	1 nF ceramic	1	(WX68Y)
C2,4	2μ2F 63 V minelect	2	(YY32K)
C3,7	22 μF 16 V minelect	2	(YY36P)
C5	100 nF 16 V minidisc	1	(YR75S)
C6	4n7F ceramic	1	(WX76H)

Semiconductors

IC1	SL6270C	1	(UL87U)
TR1	2SC2547E	1	(QY11M)

Miscellaneous

P1–10	pin 2145	1	(FL24B)
	8-pin DIL socket	1	(BL17T)

PCB	1 (GH00A)
quickstick pads	1 (HB22Y)
PP3 clip	1 (HF28F)
min screened cable	1 (XR15R)
submin omni mic insert	1 (FS43W)
instruction leaflet	1 (XT26D)
constructors' guide	1 (XH79L)

Optional (not in kit)

PP3-sized alkaline battery	1 (FK67K)

The above parts (excluding optionals) are available as a kit, order as LP98G

SSM2120 dynamic range processor

The SSM2120 is a versatile integrated circuit designed for the purpose of processing dynamic signals in various analogue systems including audio. This dynamic range processor consists of two voltage controlled amplifiers and two level detectors. These circuit blocks allow the user to logarithmically control the gain or attenuation of the signals presented to the level detectors depending on their magnitudes. This allows the compression, expansion or limiting of a.c. signals which are some of the primary applications for the SSM2120. The device will operate over a wide range of power supply voltages between ±5 V and ±18 V. Figure 3.20 shows the integrated circuit pin-out and Table 3.8 shows some typical electrical characteristics for the device. Figure 3.21 shows the integrated circuit block diagram.

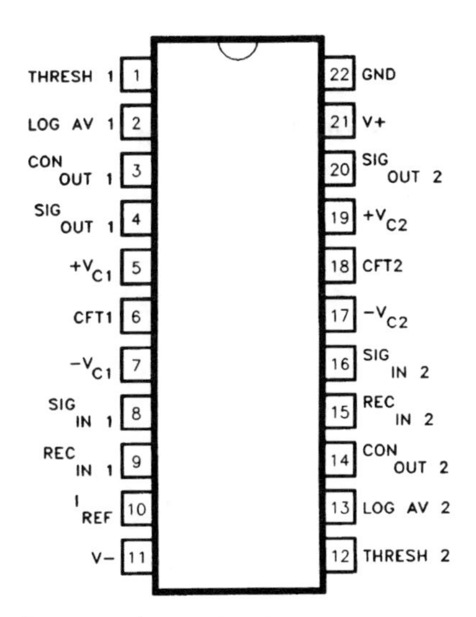

Figure 3.20 Integrated circuit pin-out

Parameter	*Conditions*	*Min*	*Typ*	*Max*
Supply voltage range				
Dual supply		±3 V	±15 V	±18 V
Single supply		+6 V	+30 V	+36 V
Supply current				
Positive			8 mA	10 mA
Negative			6 mA	8 mA
Level detectors				
Dynamic range		100 dB	110 dB	
Input current range		30 nA_{P-P}		3 mA_{P-P}
Output offset voltage			±0.5 mV	±2 mV
Frequency response				
	I_{IN} = 1 mA_{P-P}			1 MHz
	I_{IN} = 10 μA_{P-P}			50 kHz
	I_{IN} = l μA_{P-P}			7.5 kHz
VCAs				
Frequency response	Unity-gain or less			250 kHz
Control feedthrough				
(trimmed)	R_{IN} = R_{OUT} = 36 kΩ,			
	A_V 0 dB to –30 dB		750 µV	
Gain control range	Unity gain	–100 dB		+40 dB
THD (unity gain)	+10 dBV IN/OUT		0.005%	0.02%
Noise (20 kHz bandwidth)	Ref: 0 dBV		–80 dB	

Table 3.8 Typical electrical characteristics

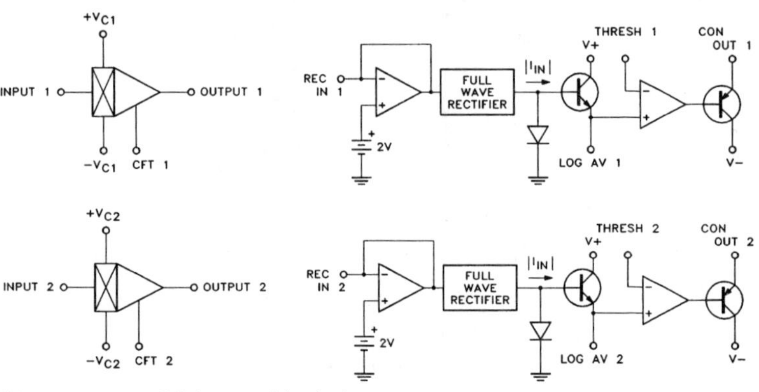

Figure 3.21 SSM2120 block diagram

Circuit description

The SSM2120 effectively contains two duplicate parts; each with a level detector and a voltage controlled amplifier.

Level detector circuit

Two independent level detection circuits are provided, each containing a wide dynamic range full-wave rectifier, logging circuit and a unipolar drive amplifier. These circuits will accurately detect the input signal level over a 100 dB range from 30 nA to 3 mA peak-to-peak.

Level detector theory of operation

Referring to the level detector block diagram of Figure 3.22, the REC_{IN} input is an a.c. virtual ground. The next block implements the full-wave rectification of the input current. This current is then fed into a logging transistor (TR1) whose pair transistor (TR2) has a fixed collector current of I_{REF}. With the use of the LOG AV capacitor, the output is then the log of the average of the absolute value of I_{IN}.

When applying signals to REC_{IN}, a blocking capacitor should be followed by an input series resistor as REC_{IN} has a d.c. offset of approximately 2.1 V above ground. Choose R_{IN} for a ±1.5 mA peak signal; for ±15 V operation, this corresponds to a value of 10 kΩ.

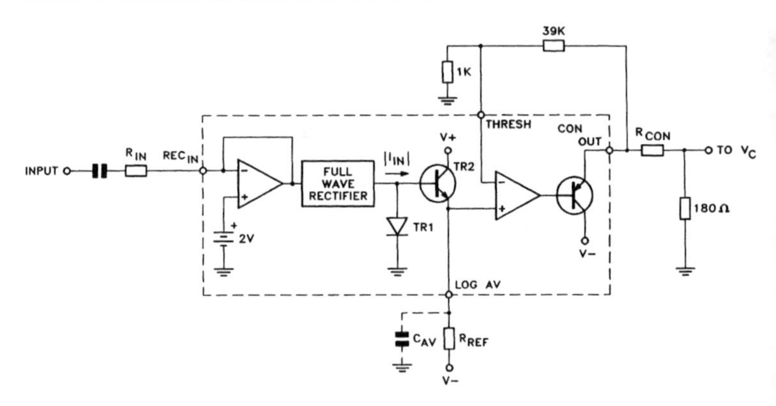

Figure 3.22 Level detector block diagram

A 1.5 MΩ value of R_{REF} from LOG AV to –15 V will establish a 10 μA reference current in the logging transistor, biasing the transistor in the middle of the detector's dynamic current range in dB to optimise dynamic range and accuracy. The LOG AV outputs are buffered and amplified by unipolar drive op-amps.

The attenuator from CON_{OUT} to the appropriate voltage controlled amplifier control port establishes the control sensitivity. Use a 180 Ω attenuator resistor to ground and choose R_{CON} for the desired sensitivity.

Voltage controlled amplifiers

The two voltage-controlled amplifiers are full Class A current in/out devices with complementary dB/V gain control ports. The control sensitivities are +6 mV/dB and –6 mV/dB. A resistor divider is used to adapt the sensitivity of an external control voltage to the range of the control port.

The signal inputs behave as virtual grounds. The input current compliance range is determined by the current into the reference current pin. This current is set by connecting a resistor to V+. The current consumption of the voltage controlled amplifiers is directly proportional to I_{REF}, which is nominally 200 μA, giving input and output clip points of ±400 μA. The device will operate at lower current levels, but with a reduced effective dynamic range.

The voltage controlled amplifier outputs are designed to interface directly with the virtual ground inputs of external operational amplifiers configured as current-to-voltage converters. The power supplies and selected compliance range determines the values of input and output resistors required. Note that the signal path through the voltage controlled amplifier, including the output current-to-voltage converter, is non-inverting.

Trimming the voltage controlled amplifiers

The control feedthrough (CFT) pins are optional control feedthrough null points. CFT nulling is required in applications such as noise gating and downward expansion. Applications such as compressors/limiters typically do not require CFT trimming because the voltage controlled amplifier operates at unity-gain, unless the signal is large enough to initiate gain reduction, in which case the signal masks control feedthrough. This trim is ineffective for voltage-controlled filter applications. If trimming is not used, leave the CFT pins open.

Kit available

A kit of parts is available to build several application circuits using the SSM2120. The kit includes a high quality fibreglass printed circuit board with a screened printed legend to aid construction, see Figure 3.23. Figure 3.24 shows the circuit diagram used to produce this printed circuit board. Note that, because the module may

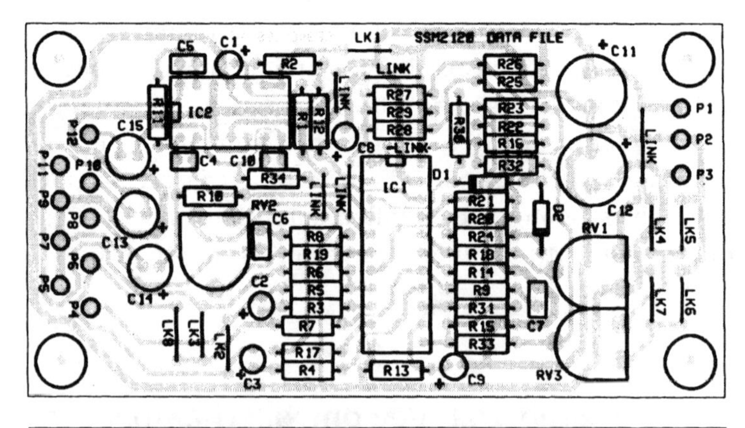

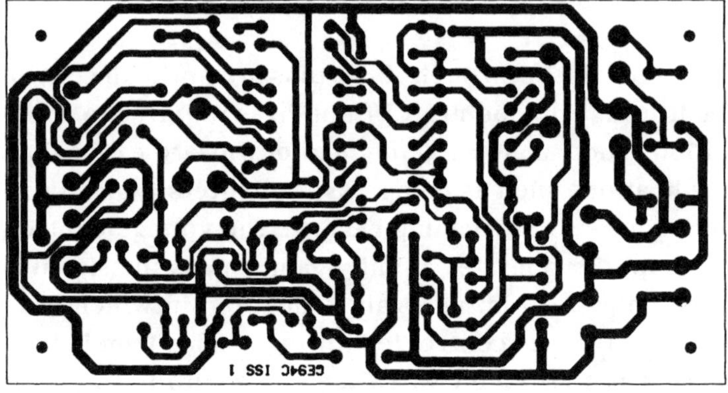

Figure 3.23 PCB legend and track

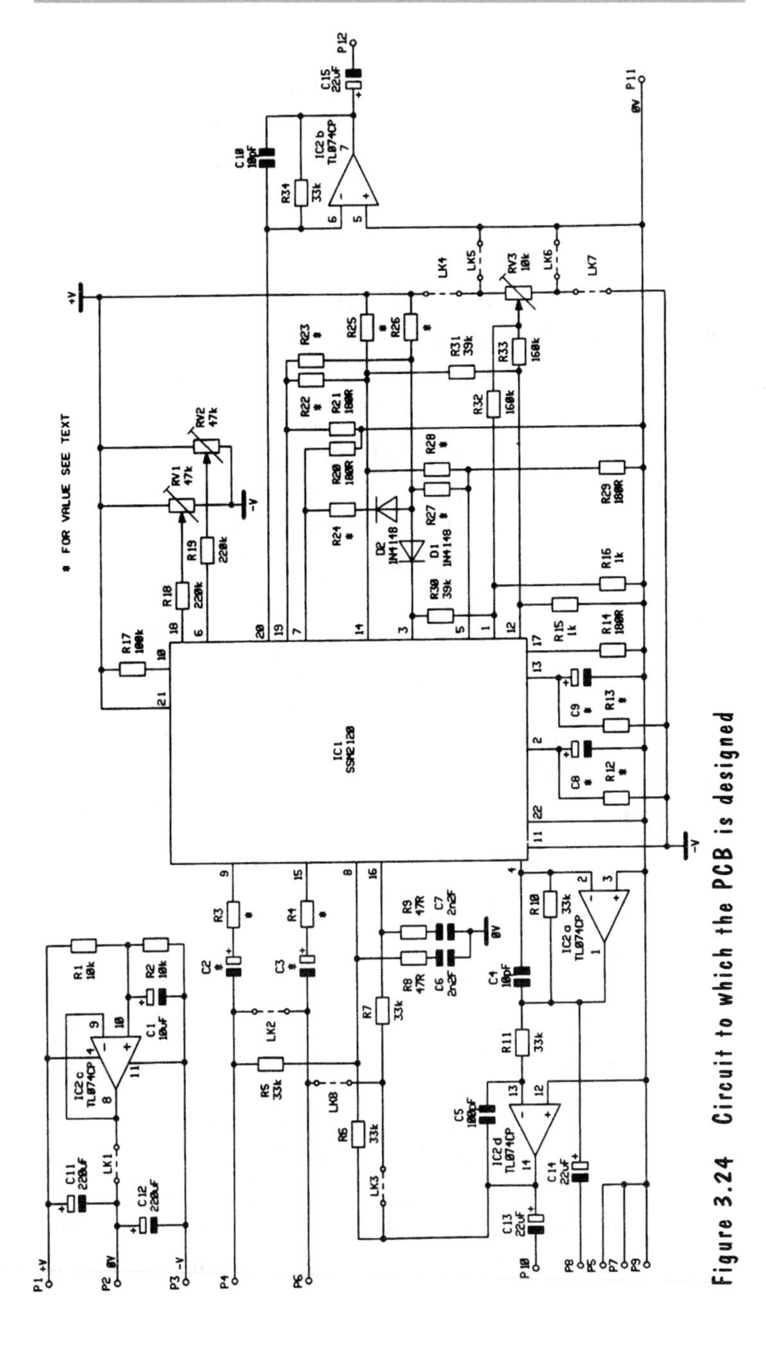

Figure 3.24 Circuit to which the PCB is designed

be used in many different applications, some of the component values supplied in the kit have been assigned an arbitrary value. For this reason minor modifications may be necessary to adapt the circuit to individual purposes.

The SSM2120 requires a split rail supply and will operate over a wide range of voltages between ±3 V and ±18 V. However, additional components have been included in the design to allow the circuit to operate from a single rail supply of between 6 V and 36 V, by installing link LK1. It is important that the supply is adequately decoupled in order to prevent the introduction of mains derived noise onto the supply rails. For optimum performance a regulated power supply should be used. All application circuits here are optimised for use with a ±15 V power supply (+30 V power supply with LK1 fitted).

The current into the reference pin determines the input and output compliance range of the voltage controlled amplifiers. This current has a nominal value of 200 μA, and is set by resistor R7; for ±15 V operation this corresponds to a value of 100 kΩ.

Figure 3.25 shows the basic wiring information.

Applications

Figure 3.26 shows the control circuit for a typical downward expander, providing a negative unipolar control output. This is typically used in downward expander, noise gate and dynamic filter applications. Here, the

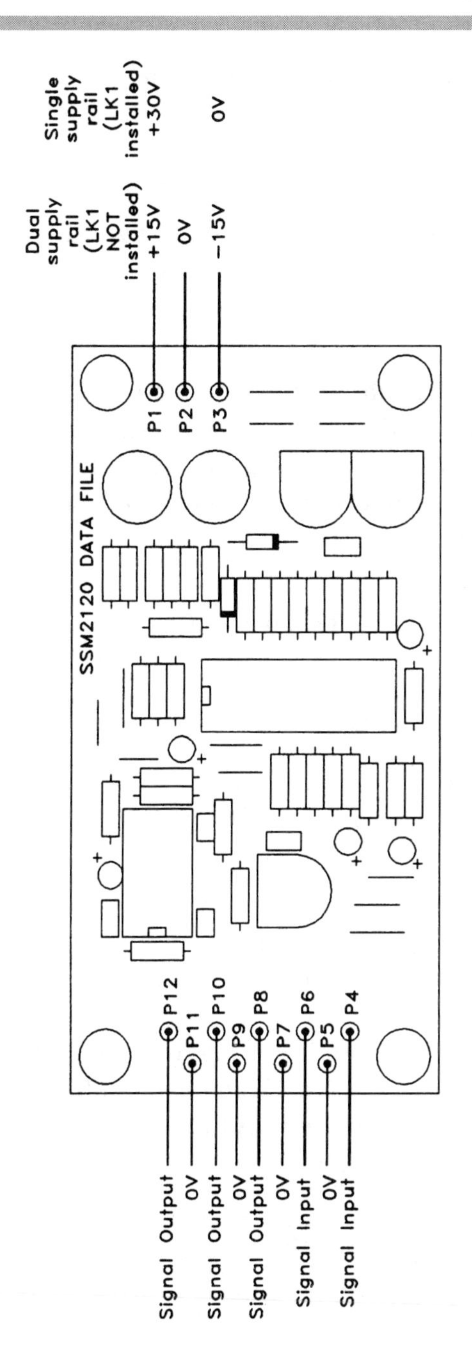

Figure 3.25 Basic wiring diagram

threshold control preset RV3 sets the signal level versus control voltage characteristics. The sensitivity of the control action depends on the value of resistor R32.

For a positive unipolar control output add two diodes, see Figure 3.27. This is useful in compressor/limiter applications.

Bipolar outputs can be achieved by connecting resistor R26 from the opamp output to V+. This is useful in compander circuits as shown in Figure 3.28. The value of resistor R26 will determine the maximum output from the control amplifier.

An attenuator resistor (R24/R27) from CON_{OUT} to the appropriate voltage controlled amplifier control port establishes the control sensitivity.

As mentioned previously, in applications such as noise gating and downward expansion the voltage controlled amplifiers require trimming as follows.

Apply a 100 Hz sinewave to the control point attenuator (D1 side of resistor R27 for voltage controlled amplifier 1, R31 side of resistor R22 for voltage controlled amplifier 2). The signal peaks should correspond to control voltages which induce the VCA's maximum intended gain and at least 30 dB of attenuation. Adjust presets RV1/RV2 for minimum feedthrough.

In all other applications, leave the CFT pins open by not fitting resistors R18 and R19.

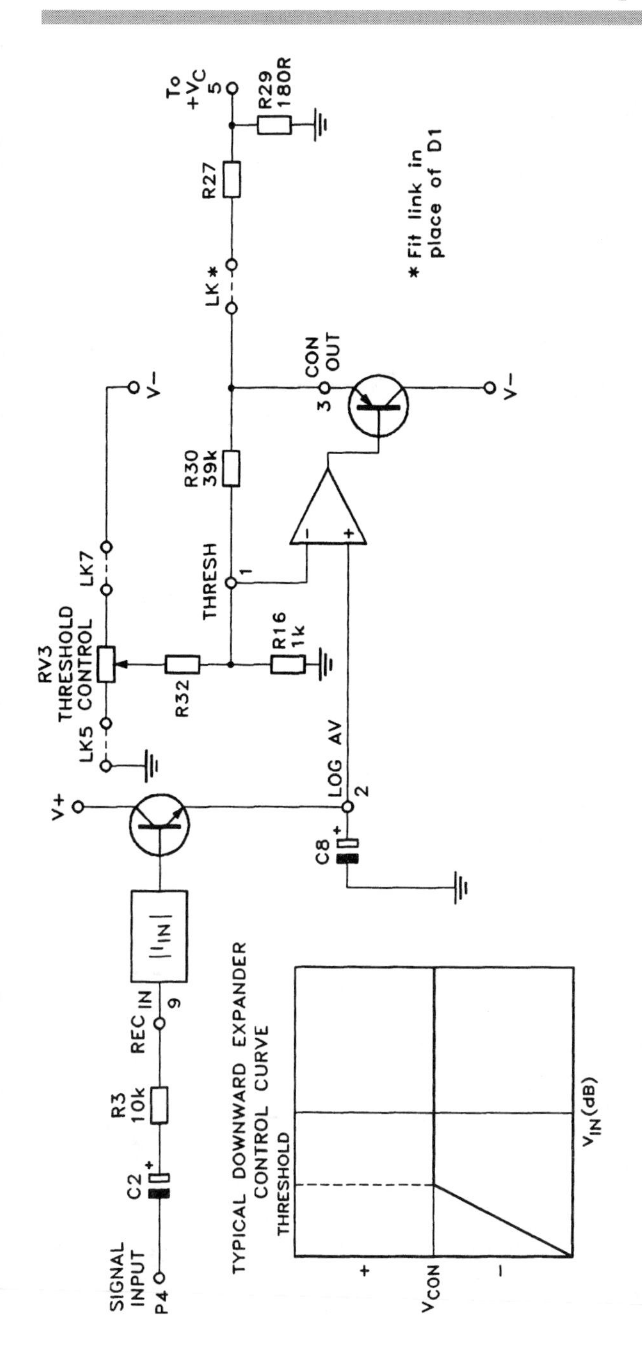

Figure 3.26 Downward expander control circuit

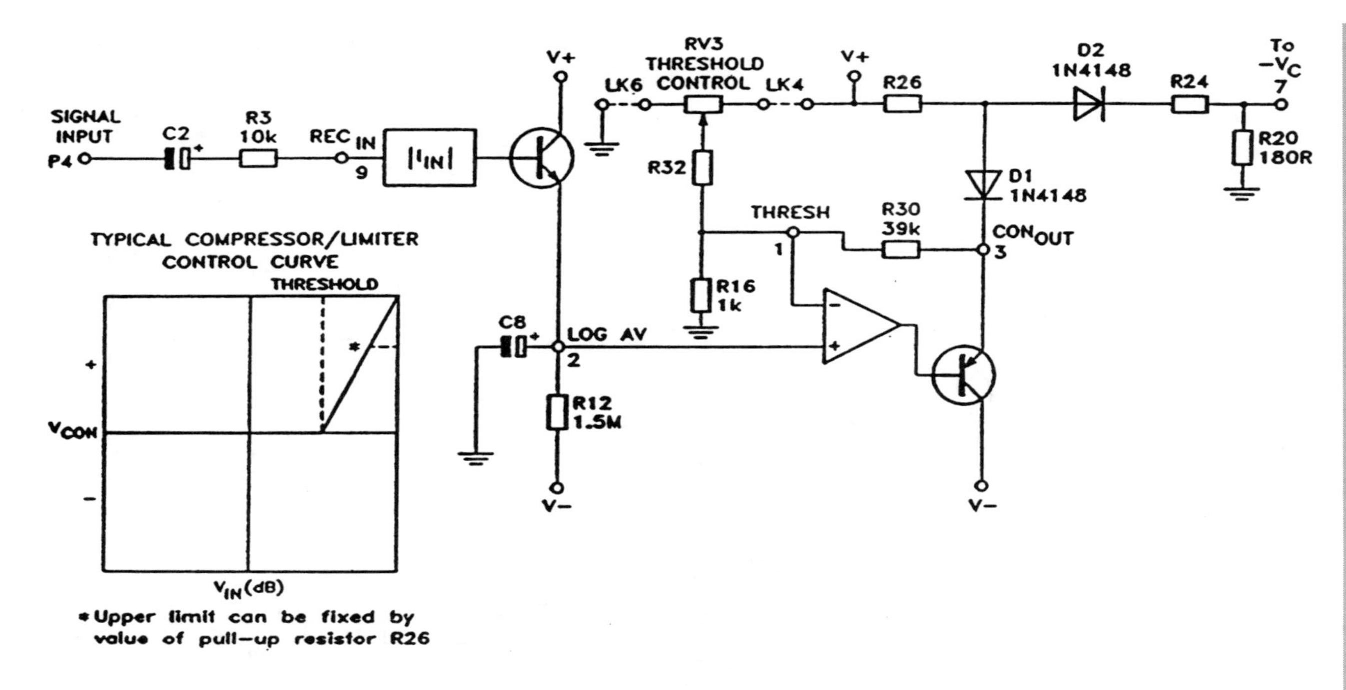

Figure 3.27 Compressor/limiter control circuit

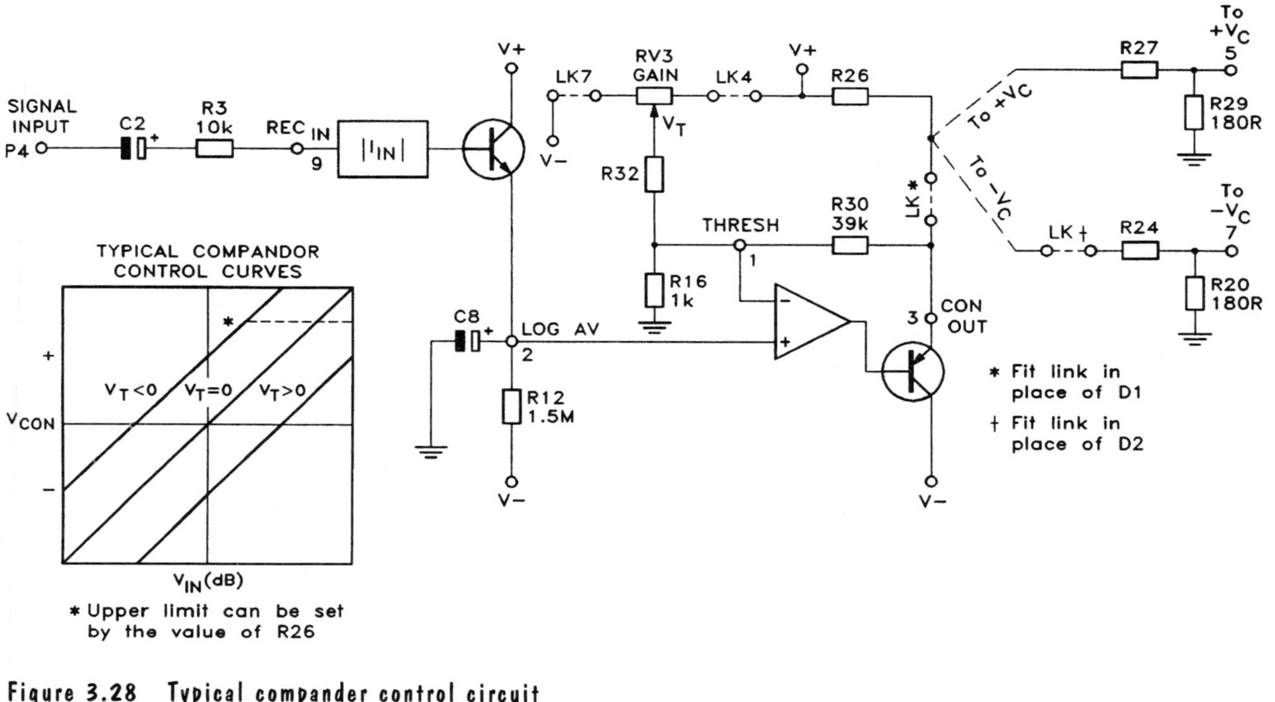

Figure 3.28 Typical compander control circuit

Companding noise reduction system

A complete companding noise reduction system is shown in Figure 3.29. Normally, to obtain an overall gain of unity, the value of resistor R24 (compression) is equal to resistor R22 (expansion). As shown in Table 3.9, the relative values of resistors R22/24 will determine the compression/expansion ratio. Note that signal compression increases gain for low level signals and reduces gain for high levels while expansion does the reverse. The exact compression/expansion ratio needed depends on the recording medium being used. For example, a household cassette deck would require a higher compression/ expansion ratio than a professional tape recorder.

Compression/ expansion/ ratio	*R22/24 (kΩ)*	*Gain (reduction or increase) (dB)*	*Compressor only output signal increase (dB)*	*Expander only output signal increase (dB)*
1.5:1	11.800	6.67	13.33	22.67
2:1	7.800	10.00	10.00	30.00
3:1	5.800	13.33	6.67	33.33
4:1	5.133	15.00	5.00	35.00
5:1	4.800	16.00	4.00	36.00
7.5:1	4.415	17.33	2.67	37.33
10:1	4.244	18.00	2.00	38.00
AGC*/limiter	3.800	20.00	0	40.00

Note: *AGC for compression only

Table 3.9 Compression/expansion ratios

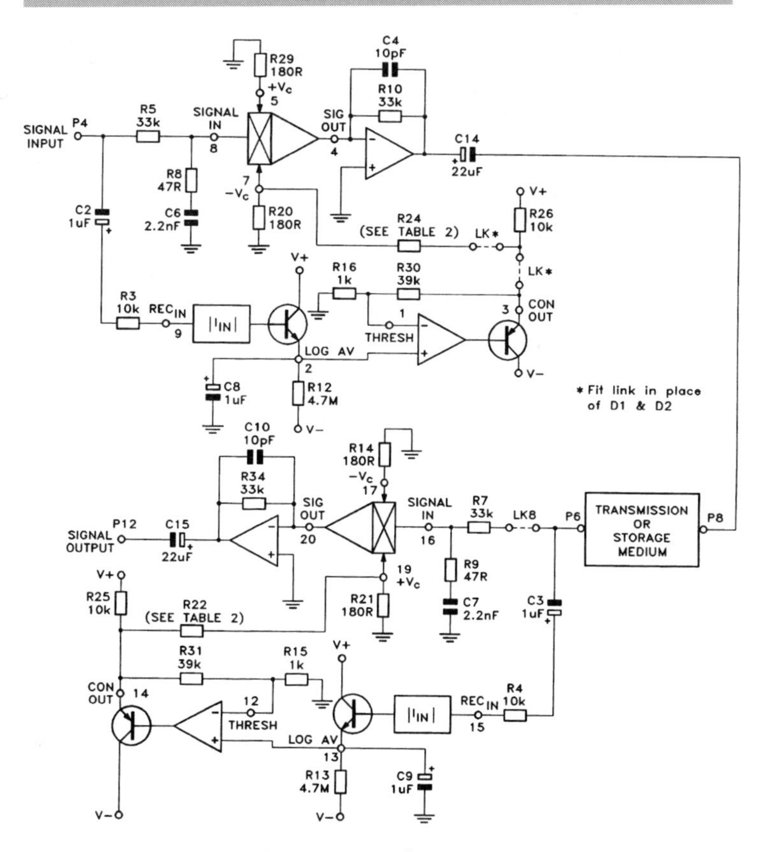

Figure 3.29 Companding noise reduction system

Dynamic filter

Figure 3.30 shows a dynamic filter capable of single ended (non-encode/decode) noise reduction. Dynamic filtering limits the signal bandwidth to less than 1 kHz unless enough *highs* are detected in the signal to cover the noise floor, when the filter opens to pass more of the audio band. Such circuits usually suffer from a loss of high-frequency content at low signal levels because their control circuits detect the absolute amount of highs present in the signal. This circuit, however, measures wideband level as well as highfrequency band level to produce a composite control signal combined in a 1:2 ratio. The upper detector senses wideband signals with a cut-off of 20 Hz, while the lower detector has a 5 kHz cut-off to sense only high-frequency band signals. Unfortunately, even in this system, a certain amount of mid- and high-frequency components will be lost, especially during transients at very low signal levels. The threshold control, preset RV3, sets the filter characteristics for 50 dB (V+) to 90 dB (V–) dynamic range programme source material.

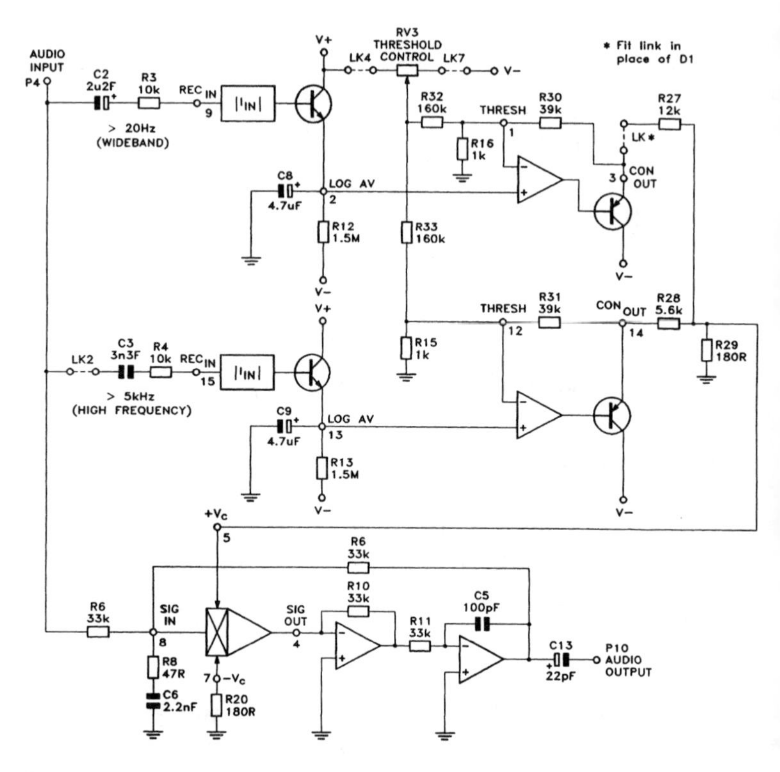

Figure 3.30 Dynamic noise filter circuit

Dynamic filter with downward expander

As shown in Figure 3.31, the output from the wideband detector can also be connected to the $+V_C$ control port of the second voltage controlled amplifier which is connected in series with the sliding filter. This will act as a downward expander with a threshold that tracks that of the filter. Downward expansion uses a voltage controlled amplifier controlled by the level detector. This section maintains dynamic range integrity for all levels

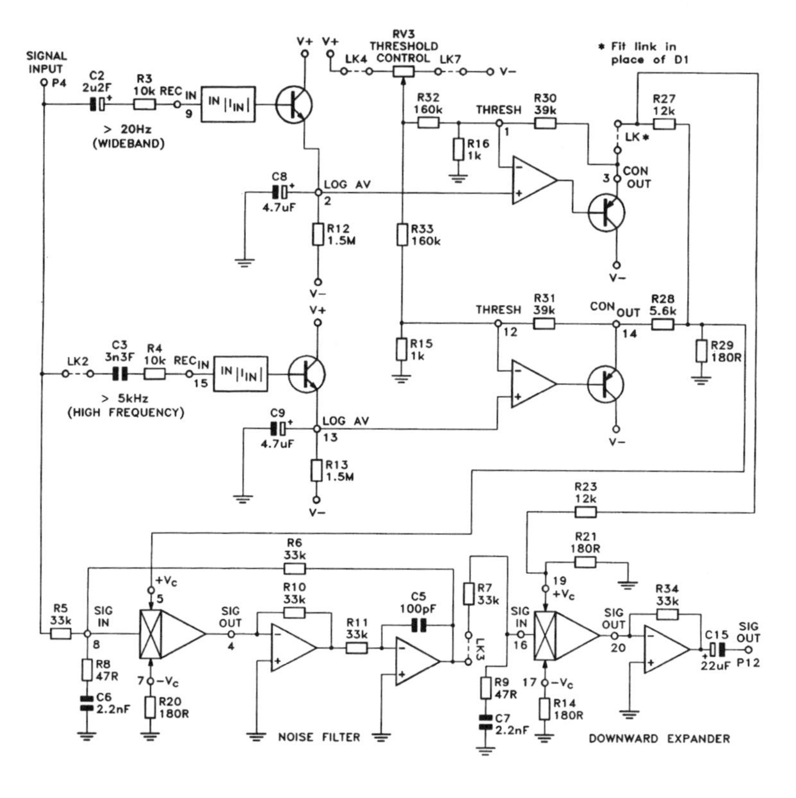

Figure 3.31 Dynamic filter with downward expander

above the threshold level (set by preset RV3) but, as the input level decreases below the threshold, gain reduction occurs at an increasing rate, as shown in Figure 3.32. This technique reduces audible noise in fadeouts or low level signal passages by keeping the standing noise floor well below the programme material. Using this system, up to 30 dB of noise reduction can be realised while preserving the crisp highs with a minimum of transient side effects.

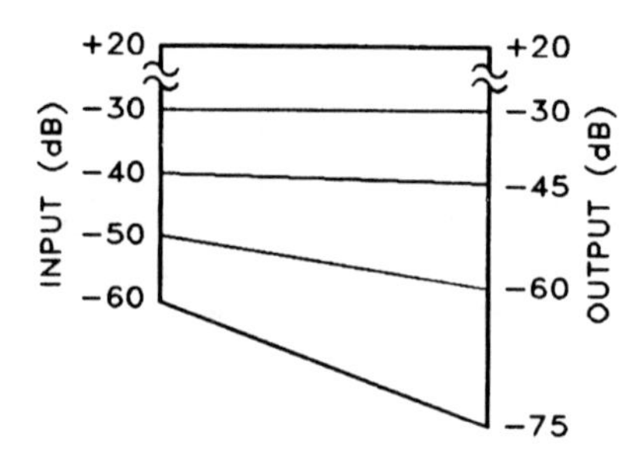

Figure 3.32 Typical downward expander I/O characteristics at –30 dB threshhold level (1:1.5 ratio)

Companding noise reduction system parts list

Resistors — All 0.6 W 1% metal film (unless specified)

R1,2,3,4, 25,26	10 k	6
R5,7,10,34	33 k	4
R8,9	47 Ω	2
R12,13	4M7	2
R14,20, 21,29	180 Ω	4
R15,16	1 k	2
R17	100 k	1
R18,19	220 k	2
R22,24	see table 2*	2
R30,31	39 k	2
RV1,2	47 k hor encl preset	2

Capacitors

C1	10 μF 16 V min elect	1
C2,3,8,9	1 μF 63 V min elect	4
C4,10	10 pF ceramic	2
C6,7	2n2F ceramic	2
C11,12	220 μF 35 V PC electrolytic	2
C14,15	22 μF 35 V min elect	2

Semiconductors

D1,2	fit link	2
IC1	SSM2120P	1
IC2	TL074CN	1

Miscellaneous

P1–12	1 mm PCB pins	12
LK8	fit link	1

*Note resistors R22 and R24 are supplied as 7k5 in the kit.

Dynamic noise filter with downward expander parts list

Resistors — All 0.6 W 1% metal film (unless specified)

R1,2,3,4	10 k	4
R5,6,7,10, 11,34	33 k	6
R8,9	47 Ω	2
R12,13	1M5	2
R14,20, 21,29	180 Ω	4
R15,16	1 k	2
R17	100 k	1
R18,19	220 k	2
R23,27	12 k	2
R28	5k6	1
R30,31	39 k	2
R32,33	160 k	2
RV1,2	47 k hor encl preset	2
RV3	10 k hor encl preset	1

Capacitors

C1	10 µF 16 V min elect	1
C2	2 µ2F 63 V min elect	1
C3	3n3F ceramic	1
C5	100 pF ceramic	1
C6,7	2n2F ceramic	2
C8,9	4 µ7F 35 V min elect	2
C11,12	220 µF 35 V PC electrolytic	2
C15	22 µF 35 V min elect	1

Semiconductors

D1	fit link	1
IC1	SSM2120P	1
IC2	TL074CN	1

Miscellaneous

P1–12	1 mm PCB pins	12
LK2,3,4,7	fit link	4

Dynamic noise filter parts list

Resistors — All 0.6 W 1% metal film (unless specified)

R1,2,3,4	10 k	4
R5,6,7,10,11	33 k	4
R8	47 Ω	1
R12,13	1M5	2

R20,29	180 Ω	2
R15,16	1 k	2
R17	100 k	1
R19	220 k	1
R27	12 k	1
R28	5k6	1
R30,31	39 k	2
R32,33	160 k	2
RV2	47 k hor encl preset	1
RV3	10 k hor pncl preset	1

Capacitors

C1	10 μF 16 V min elect	1
C2	2μ2F 63 V min elect	1
C3	3n3F ceramic	1
C5	100 pF ceramic	1
C6	2n2F ceramic	1
C8,9	4 μ7F 35 V min elect	2
C11,12	220 μF 35 V PC electrolytic	2
C13	22 μF 35 V min elect	1

Semiconductors

D1	fit link	1
IC1	SSM2120P	1
IC2	TL074CN	1

Miscellaneous

P1–12	1 mm PCB pins	12
LK2,4,7	fit link	3
Magazine Version		

SSM2120 parts list

Resistors — All 0.6 W 1% metal film (unless specified)

R1,2,3,4, 25,26	10 k	6	(M10K)
R5,6,7, 10,11.34	33 k	6	(M33K)
R8,9	47 Ω	2	(M47R)
R12,13	4M7	2	(M4M7)
R12,13	1M5	2	(M1M5)
R14,20,21,29	180 Ω	4	(M180R)
R15,16	1 k	2	(M1K)
R17	100 k	1	(M100K)
R18,19	220 k	2	(M220K)
R22,24	see table 2 (nominally 7k5)	2	(M7K5)
R23,27	12 k	2	(M12K)
R28	5k6	1	(M5K6)
R30,31	39 k	2	(M39K)
R32,33	160 k	2	(M160K)
RV1,2	47 k hor encl preset	2	(UH05F)
RV3	10 k hor encl preset	1	(UH03D)

Capacitors

C1	10 μF 16 V min elect	1	(YY34M)
C2,3,8,9	1 μF 63 V min elect	4	(YY31J)
C2	2μ2F 63 V min elect	1	(YY32K)
C3	3n3F ceramic	1	(WX74R)
C4,10	10 pF ceramic	2	(WX44X)
C5	100 pF ceramic	1	(WX56L)

C6,7	2n2F ceramic	2 (WX72P)
C8,9	4μ7F 35 V min elect	2 (YY33L)
C11,12	220 μF 35 V PC electrolytic	2 (JL22Y)
C13,14,15	22 μF 35 V min elect	3 (RA54D)

Semiconductors

D1,2	1N4148	2 (QL80B)
IC1	SSM2120P	1 (UL78K)
IC2	TL074CN	1 (RA69A)

Miscellaneous

P1–12	1 mm PCB pins	1 (FL24B)
	printed circuit board	1 (GE94C)
	constructors' guide	1 (XH79L)

4 Miscellaneous project

Featuring:

UA3730 electronic code-lock

The UA3730 is a single chip electronic code-lock integrated circuit utilising CMOS technology. The IC can handle passcodes of up to 12 digits, allowing the use of up to 1 million million unique codes. If an incorrect code is entered three times or more, an output is activated allowing a burglar alarm to be triggered. The code may be changed as many times as required by the user providing additional security. Figure 4.1 shows the integrated circuit pin-out, and Table 4.1 lists typical electrical characteristics for the device. Also please note that the UA3730 may also be correctly supplied and marked as SH901. Figure 4.2 shows a typical application circuit for the device. The integrated circuit requires a battery back-up to prevent the code from being lost if the main power supply fails. When the device is initially powered

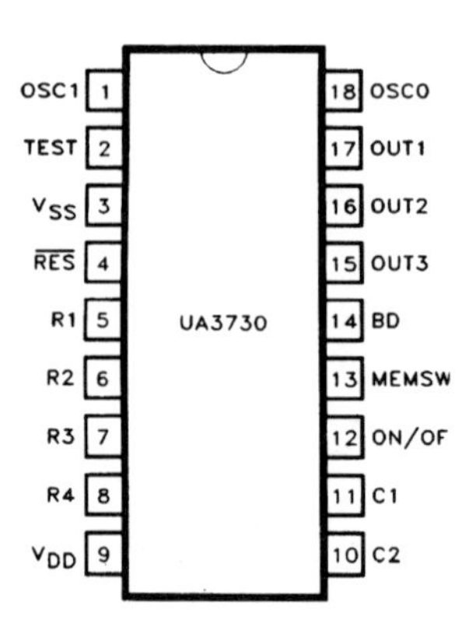

Figure 4.1 IC pin-out

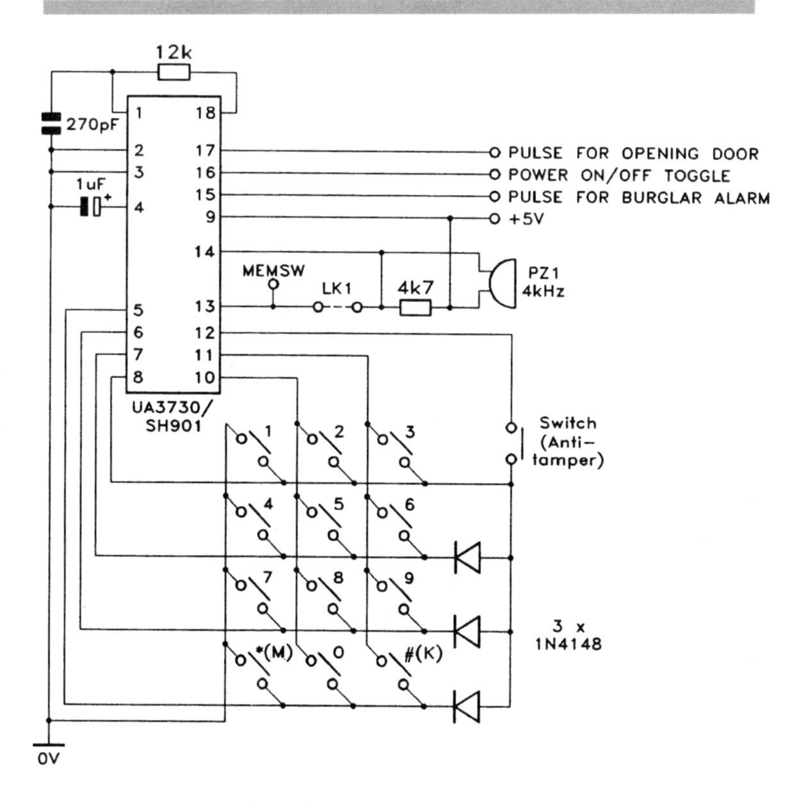

Figure 4.2 Typical application circuit

up, the code is automatically set to 0. If IC pin 13 is connected to 0 V, the device is set to the program mode. A new code may then be set as follows: key in the new code (up to 12 digits) followed by *M*. The code has then been changed to that entered. After the new code has been set, pin 13 should be disconnected from 0 V and returned to the *floating* state, preventing unauthorised users keying in further new codes.

Parameter	Conditions	Min	Typ	Max
Power supply voltage		3 V	5 V	6 V
Operating current	Operating frequency = 400 kHz	1 mA		4 mA
Stand-by current			5 μA	30 μA
Operating frequency		260 kHz	400 kHz	645 kHz
Output current (OUT1–3)				15 mA
Storage temperature		–55°C		125°C
Operating temperature		–30°C		70°C

Table 4.1 Typical electrical characteristics

The circuit may also be set to the programming mode using the keypad, if pin 13 and pin 14 of the integrated circuit are linked together. In this case the code may be changed as follows: key in the current code (which is set to *0* when the unit is powered up) followed by *M*. The circuit is now set to the programming mode. Then key in the new code followed by *M*. The new code is then set and the circuit returns to the previous state.

There are three outputs which are activated from the keypad. OUT1 activates for 2 seconds when the correct code followed by K(#) is entered. OUT2 changes state when the correct code followed by K is entered providing a toggle action. OUT3 activates for one minute if an incorrect code is entered three times or more, and this output may be used to trigger an alarm. All outputs are open drain types and require a pullup resistor. Integrated circuit pins 12 and 8 may also be used to trigger an alarm condition, and may be connected to a mechanical microswitch to provide additional anti-tamper protection.

Kit available

A kit of parts is available, including a fibreglass printed circuit board with a screen printed legend, for a basic application circuit using the UA3730. A simple numeric keypad is also supplied in the kit. Figure 4.3 shows the circuit diagram of the module, and Figure 4.4 shows the printed circuit board legend. The printed circuit board is designed to mount onto the rear of the keypad and be wired directly to the terminals. Figure 4.5 shows how to mount the printed circuit board onto the keypad using M3 spacers and washers and M2.5 nuts and bolts. The keypad is connected to the printed circuit board using 24 SWG tinned copper wire as shown in Figure 4.6. Other input and output connections to the module are made via double-sided printed circuit board pins, and may be taken from either side of the board. The pins are fitted such that the thin end of the pin is inserted into the printed circuit board from the track side. A different keypad may be wired onto the board in place of the keypad supplied as long as the matrix is correct. The required matrix is shown in Figure 4.7. A piezo sounder is included in the design as an aid to the user. The length of the output from the sounder indicates whether or not the input has been accepted, as shown in Table 4.2. LEDs may also be connected to indicate output status.

The module requires a regulated 4 V to 6.5 V power supply that is capable of supplying at least 20 mA. The power supply should be well decoupled to prevent the introduction of noise onto the supply rail, as this could produce unpredictable results. If the power supply is also used to drive an output load connected to the mod-

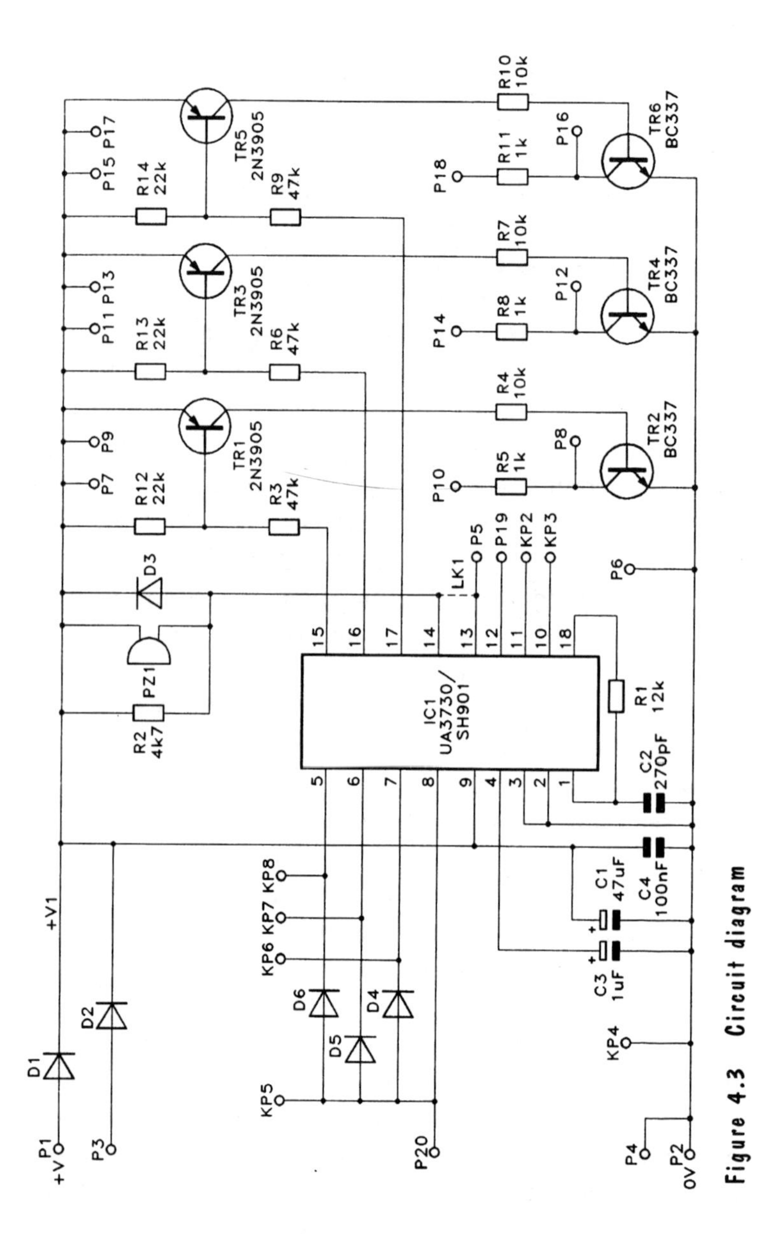

Figure 4.3 Circuit diagram

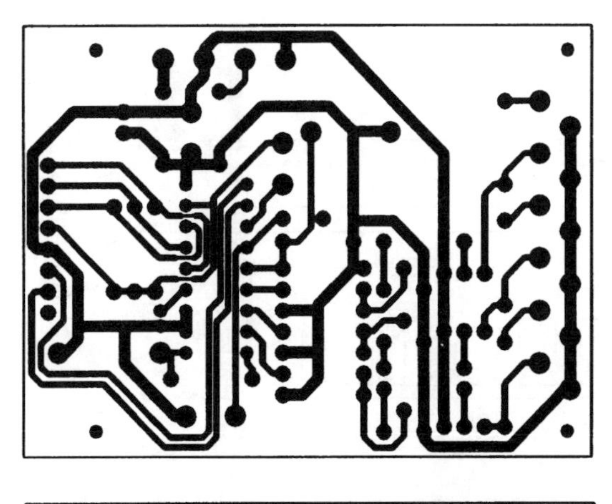

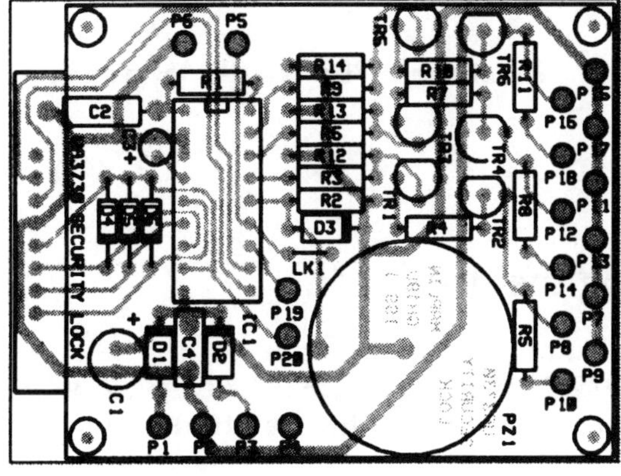

Figure 4.4 PCB legend

ule, then the total current consumption of the load must also be added to the power supply capacity; for example, if the total load current is 30 mA, then the power supply must be capable of supplying at least 50 mA (20 mA for the module and LEDs + 30 mA total load current). Figure 4.8 shows a simple 5 V regulator circuit suitable to power just the module and LEDs (any output loads should be connected to a separate supply).

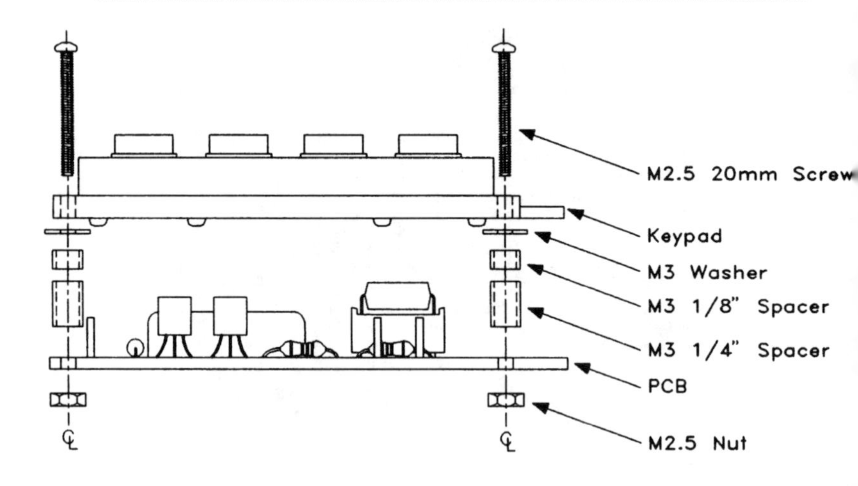

Figure 4.5 Mounting the PCB onto the keypad

Setting the user code

There are two methods of setting a user code as detailed above. Link LK1 may either be fitted or omitted depending on the chosen method of setting the code. The options are as follows.

● LK1 not fitted

To set the new code, link P5 to P6 and enter the new code (up to 12 digits) followed by the * symbol on the keypad. The new code is then set and the circuit will no longer respond to any previous codes. After the new code is set, the link between P5 and P6 should then be removed.

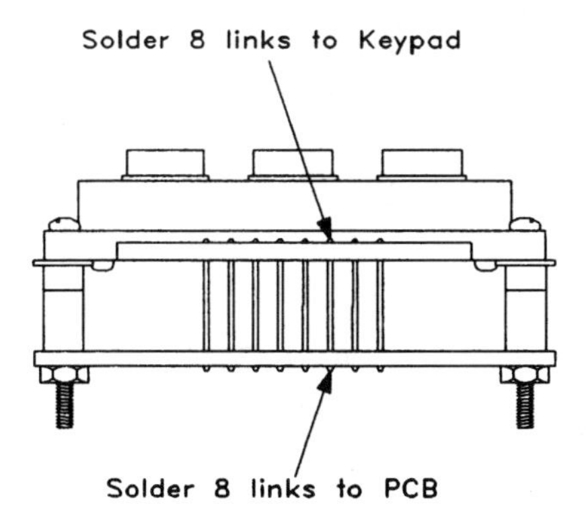

Figure 4.6 Linking the keypad to the PCB

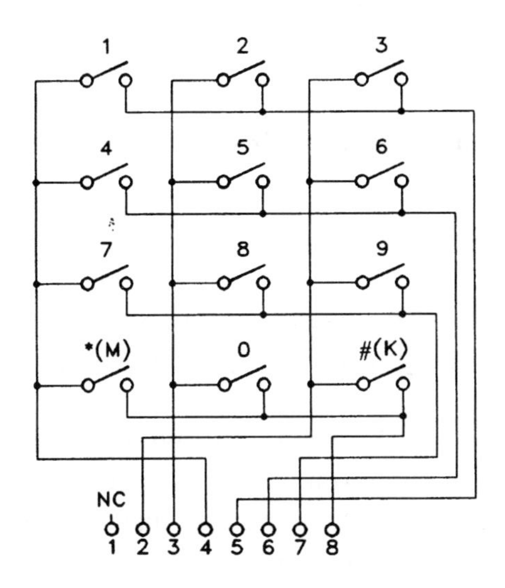

Figure 4.7 Keypad matrix

Condition	Number of pulses from sounder	Outputs P16	P12	P8
		OUT1	OUT2	OUT3
Input correct code	1	Active for 2s	Toggle	
1st and 2nd wrong inputs	2			
3rd wrong input; Tamper (on/off) input	Groups of 3 pulses for 1 minute			
P19 connected to P20	Active for 1 minute			

Table 4.2 Output conditions

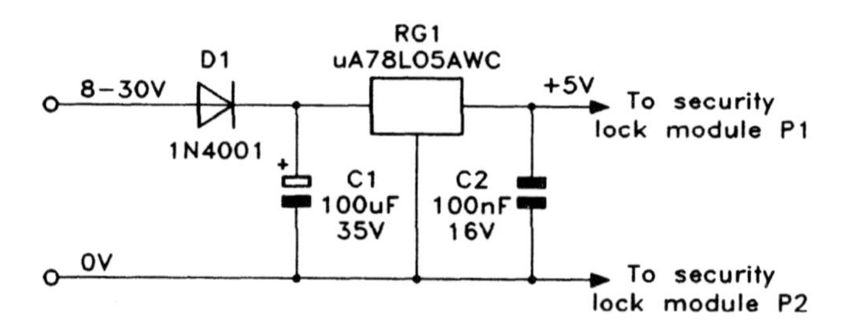

Figure 4.8 Simple regulator circuit

● LK1 fitted

To set the new code, key in the current code (*0* when the circuit is initially powered up) followed by the * symbol; the circuit is now in the program mode. Enter the new code (up to 12 digits) followed by *. The new code is then set and the circuit will no longer respond to any previous codes. This procedure must be followed each time a new code is set. If LK1 is fitted P5 and P6 must *not* be linked at any time. The * button on the keypad is only used when a new code is being set.

For normal use, the code is entered to activate and deactivate the outputs to open and close doors, switch lights on and off, and so on. To operate the outputs, key in the code (which you have previously set using the procedure above) followed by the # symbol on the keypad.

Battery back-up

If the user wishes to retain the code when the power supply is removed, then a back-up battery supply must be provided. Facility for the connection of a back-up battery is provided by P3(battery +) and P4(0 V). The voltage of the main supply should be kept at least 0.5 V above the back-up battery voltage to prevent the battery being drained in normal use. The battery voltage should be between 3.7 and 6 V. Typically, with a main supply voltage of between 5 V and 6 V, a back-up 4 to 4.5 V battery pack could be used. The quiescent current of the circuit in stand-by mode is typically in the order of a few hundred nA and so the drain on even a low capacity back-up battery is very little when the circuit is in the quiescent state.

Alarm condition

If an incorrect code is entered more than twice then the circuit latches into an alarm condition and the sounder pulses for 1 minute or until the correct code is entered.

An alarm condition can also be created by connecting P19 to P20 using a mechanical switch; a microswitch may be connected to this input to provide a degree of anti-tamper protection. It should be noted that this input may be used to activate the alarm output (OUT3) independently of the previous state of the circuit. In addition to triggering an alarm, the alarm output may also be used to trigger additional circuitry, providing a *lockout* function. This means access is denied for a set period after the alarm output has been triggered, even if the correct code is subsequently entered and the keypad disabled. Figure 4.9 shows an idea for a simple *lockout* circuit, featuring a lockout time of around 5 minutes.

Outputs are provided for 3 LED indicators to show the state of the outputs OUT1 to OUT3. In some cases, the user may not wish to fit the LEDs as they do add to the current drain. If a visual indication is not required, then it is quite acceptable to omit the LEDs.

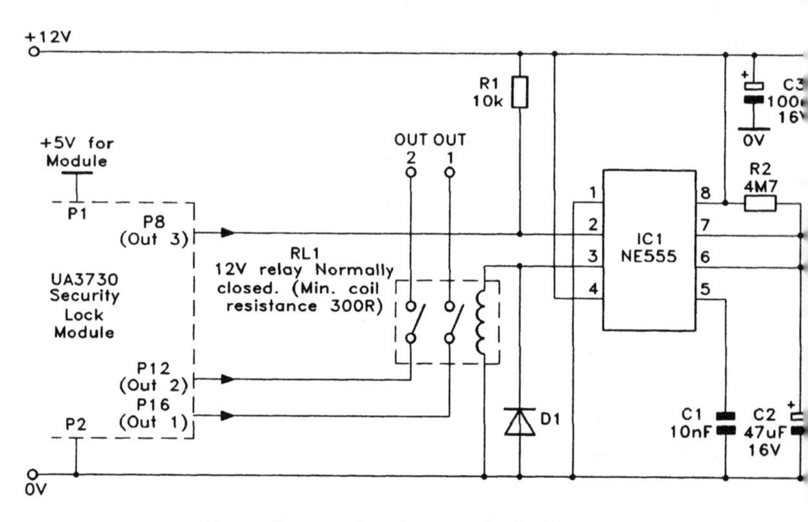

Figure 4.9 Typical example of a simple lockout circuit

The three outputs from the module are of the open collector type and the load is connected between the appropriate output pin (P8, P12 or P16) and either a +V1 pin on the printed circuit board (P7, P11 or P15) or another suitable +V supply (15 V maximum). The load current should not be allowed to exceed 40 mA at any time, as irreparable damage could occur. Finally, Table 4.3 shows the specification of the prototype UA3730 security lock module.

Figure 4.10 shows the wiring diagram for the module, showing how to connect the LEDs, back-up battery, and so on.

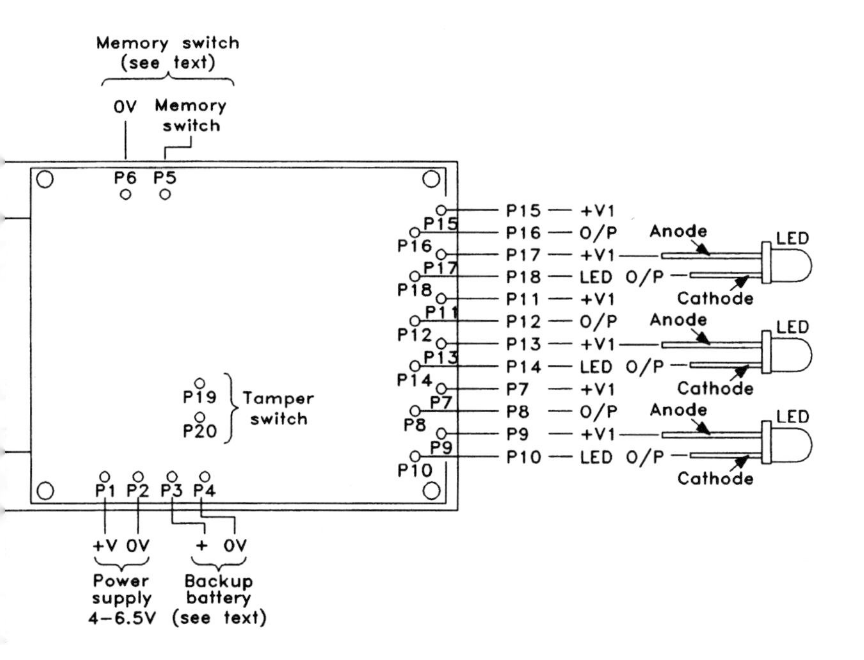

Figure 4.10 Wiring diagram

Parameter	Condition	Typical
Power supply voltage		4 V to 6.5 V
Back-up battery voltage	< supply voltage –0.5 V	3.7 V to 6 V
Power supply current	Stand-by	220 nA at 5 V
	Operating (without LEDs)	310 μA at 5 V
	Operating (LEDs active)	17 mA at 5 V
PCB size		51 mm x 71 mm
Outputs	Open collector type	
Output sink current	Absolute maximum	40 mA

Table 4.3 Specification of prototype

Electronic code-lock parts list

Resistors — All 0.6 W 1% metal film

R1	12 k	1	(M12K)
R2	4k7	1	(M4K7)
R3,6,9	47 k	3	(M47K)
R4,7,10	10 k	3	(M10K)
R5,8,11	1 k	3	(M1K)
R12,13,14	22 k	3	(M22K)

Capacitors

C1	47 μF 16 V minelect	1	(YY37S)
C2	270 pF 1% polystyrene	1	(BX50E)
C3	1 μF 63 V minelect	1	(YY31J)
C4	100 nF 16 V minidisc	1	(YR75S)

Semiconductors

IC1	UA3730	1	(UM98G)
D1–3	1N4001	3	(QL73Q)
D4–6	1N4148	3	(QL80B)
TR1,3,5	2N3905	3	(QR41U)
TR2,4,6	BC337	3	(QB68Y)

Miscellaneous

	18-pin DIL socket	1	(HQ76H)
	numeric keypad	1	(JM09K)
	PCB piezo sounder	1	(JH24B)

pin 2144	1	(FL23A)
M2.5 x 20 mm steel screw	1	(JY32K)
M2.5 steel nut	1	(JD62S)
M3 x 1/8 in spacer	1	(FG32K)
M3 x 1/4 in spacer	1	(FG33L)
M3 steel washer	1	(JD76H)
24 SWG 0.56 mm TC wire	1	(BL15R)
PCB	1	(GH18U)
instruction leaflet	1	(XT75S)
constructors' guide	1	(XH79L)

Optional (not in kit)

5 mm 2 mA LED red	1	(UK48C)
5 mm 2 mA LED green	1	(UK49D)
5 mm 2 mA LED yellow	1	(UK50E)

The above parts (excluding optionals) are available as a kit, order as LP92A